EAST EUROPEAN ACADEMIES IN TRANSITION

Sociology of the Sciences Library

1. R. Mayntz, U. Schimank and P. Weingart (eds.): *East European Academies in Transition*. 1998
ISBN 0-7923-5168-1

Published in the Sociology of the Sciences Monographs

Marc de Mey, *The Cognitive Paradigm*, 1982, xx + 314 pp., ISBN 90-277-1382-0.
Tom Jagtenberg, *The Social Constructions of Science*, 1983 xviii + 237 pp., ISBN 90-277-1498-3.
Norman Stockman, *Antipositivist Theories of the Sciences*, 1983, x+ 284 pp., ISBN 90-277-1567-X.
Rachel Laudan (ed.), *The Nature of Technological Knowledge. Are Models of Scientific Change Relevant?* 1984, vii + 145 pp., ISBN 90-277-1716-8.
Trevor Pinch, *Confronting Nature*, 1986, xi + 268 pp., ISBN 90-277-2224-2.
Olga Amsterdamska, *Schools of Thought. The Development of Linguistics from Bopp to Saussure*, 1987, x + 320 pp., ISBN 90-277–2391-5

Published in the Sociology of the Sciences Yearbook

1. E. Mendelsohn, P. Weingart and R. Whitley (eds.): *The Social Production of Scientific Knowledge*. 1977 ISBN Hb 90-277-0775-8; Pb 90-277-0776-6
2. W. Krohn, E.T. Layton, Jr. and P. Weingart (eds.): *The Dynamics of Science and Technology*. Social Values, Technical Norms and Scientific Criteria in the Development of Knowledge. 1978 ISBN Hb 90-277-0880-0; Pb 90-277-0881-9
3. H. Nowotny and H. Rose (eds.): *Counter-Movements in the Sciences*. The Sociology of the Alternatives to Big Science. 1979 ISBN Hb 90-277-0971-8; Pb 90-277-0972-6
4. K.D. Knorr, R. Krohn and R. Whitley (eds.): *The Social Process of Scientific Investigation*. 1980 (1981) ISBN Hb 90-277-1174-7; Pb 90-277-1175-5
5. E. Mendelsohn and Y. Elkana (eds.): *Sciences and Cultures*. Anthropological and Historical Studies of the Sciences. 1981 ISBN Hb 90-277-1234-4; Pb 90-277-1235-2
6. N. Elias, H. Martins and R. Whitley (eds.): *Scientific Establishments and Hierarchies*. 1982 ISBN Hb 90-277-1322-7; Pb 90-277-1323-5
7. L. Graham, W. Lepenies and P. Weingart (eds.): *Functions and Uses of Disciplinary Histories*. 1983 ISBN Hb 90-277-1520-3; Pb 90-277-1521-1
8. E. Mendelsohn and H. Nowotny (eds.): *Nineteen Eighty Four: Science between Utopia and Dystopia*. 1984 ISBN Hb 90-277-1719-2; Pb 90-277-1721-4
9. T. Shinn and R. Whitley (eds.): *Expository Science*. Forms and Functions of Popularisation. 1985 ISBN Hb 90-277-1831-8; Pb 90-277-1832-6
10. G. Böhme and N. Stehr (eds.): *The Knowledge Society*. The Growing Impact of Scientific Knowledge on Social Relations. 1986 ISBN Hb 90-277-2305-2; Pb 90-277-2306-0
11. S. Blume, J. Bunders, L. Leydesdorff and R. Whitley (eds.): *The Social Direction of the Public Sciences*. Causes and Consequences of Co-operation between Scientists and Non-scientific Groups. 1987 ISBN Hb 90-277-2381-8; Pb 90-277-2382-6
12. E. Mendelsohn, M.R. Smith and P. Weingart (eds.): *Science, Technology and the Military*. 2 vols. 1988 ISBN Vol, 12/1 90-277-2780-5; Vol. 12/2 90-277-2783-X

Sociology of the Sciences Library

---

13. S. Fuller, M. de Mey, T. Shinn and S. Woolgar (eds.): *The Cognitive Turn*. Sociological and Psychological Perspectives on Science. 1989 ISBN 0-7923-0306-7
14. W. Krohn, G. Kp̈pers and H. Nowotny (eds.): *Selforganization*. Portrait of a Scientific Revolution. 1990 ISBN 0-7923-0830-1
15. P. Wagner, B. Wittrock and R. Whitley (eds.): *Discourses on Society*. The Shaping on the Social Science Disciplines. 1991 ISBN 0-7923-1001-2
16. E. Crawford, T. Shinn and S. Sr̈lin (eds.): *Denationalizing Science*. The Contexts of International Scientific Practice. 1992 (1993) ISBN 0-7923-1855-2
17. Y. Ezrahi, E. Mendelsohn and H. Segal (eds.): *Technology, Pessimism, and Postmodernism*. 1993 (1994) ISBN 0-7923-2630-X
18. S. Maasen, E. Mendelsohn and P. Weingart (eds.): *Biology as Society? Society as Biology: Metaphors*. 1994 (1995) ISBN 0-7923-3174-5
19. T. Shinn, J. Spaapen and V. Krishna (eds.): *Science and Technology in a Developing World*. 1995 (1997) ISBN 0-7923-4419-7
20. J. Heilbron, L. Magnusson and B.Wittrock (eds.): *The Rise of the Social Sciences and the Formation of Modernity*. Conceptual Change in Context, 1750–1850. 1996 (1998) ISBN Hb 90-277-1234-4; Pb 90-277-1235-2

East European Academies in Transition

Edited by

RENATE MAYNTZ

Max-Planck-Institut für Gesellschaftsforschung, Köln, Germany

UWE SCHIMANK

Fernuniversität Hagen, Germany

and

PETER WEINGART

Universität Bielefeld, Germany

KLUWER ACADEMIC PUBLISHERS

DORDRECHT / BOSTON / LONDON

A C.I.P. Catalogue record for this book is available from the Library of Congress.

ISBN 0-7923-5168-1

Published by Kluwer Academic Publishers,
P.O. Box 17, 3300 AA Dordrecht, The Netherlands.

Sold and distributed in North, Central and South America
by Kluwer Academic Publishers,
101 Philip Drive, Norwell, MA 02061, U.S.A.

In all other countries, sold and distributed
by Kluwer Academic Publishers,
P.O. Box 322, 3300 AH Dordrecht, The Netherlands.

Printed on acid-free paper

Printed in the Netherlands.

TABLE OF CONTENTS

Introduction ix

The Impact of Radical Regime Change on the East European Academies of Sciences 1
Renate Mayntz

Financial Threats to Organizational Survival 13
Julita Jablecka

Coping with Scarcity: Strategies of Personnel Reduction 33
Stanislav Provaznik/Petr Machleidt/Adolf Filácek

Identity Problems: Basic or Applied Research? 49
György Darvas

What Determined an Institute's Fate in Transformation? 63
Uwe Schimank

Antagonistic Cooperation in Academy-University Relations 79
Eduard Sarmir/Stefan Zajac

Compromised Futures: The Consequences of an Aging Research Staff 93
Gennady A. Nesvetailov

From East to West: New Patterns of International Relations of Research 107
Elena Mirskaya

The Two-Edged Sword of Autonomy: Changes in the Academy-Institute Relations 125
Kostadinka Simeonova

Self-Conception, Strategies of Transformation, and Moods in Coping 141
Peter Weingart

Appendix: 151

Subject Index 157

Author Index 163

List of Contributors 165

INTRODUCTION

When the socialist regime in Central and Eastern Europe (CEE) was overthrown around the end of the eighties, beginning of the nineties, an overall transformation of whole societies started. Not only the political and the economic systems of these countries, but all societal sectors underwent deep changes. These changes presented opportunities, but they also spelled trouble. On one hand, getting rid of stifling political control and excessive bureaucratic regulation was something which most members of these societies desired. On the other, it became apparent very soon that the necessary and long hoped-for rebuilding of the economy, education, health care, the mass media, and science, too, was strongly restricted by the scarcity of financial resources. After a short period, during which opportunities were energetically taken up in a spirit of hope, came a long and still lasting time of growing troubles and despondency. Only in a few of the CEE countries have some glimpses of hope become visible recently; and it remains to be seen whether these signals are reliable.

Until now, therefore, the transformation dynamics of all societal sectors in all of the CEE countries have primarily been troublesome. This is surely true for the post-socialist research systems.[1] The demise of the communist party's absolute rule over society has allowed researchers the public expression and the pursuit of goals whose common denominator has been a greater self-regulation of scientific research according to its own criteria and logic. Legally, the autonomy of science – on the level of individual researchers as well as institutes – as it is known in the West was institutionalized in all countries. But economic crisis has deepened, and this has brought about a dramatic fiscal crisis of the state. In this situation, the political priorities for spending these reduced financial resources have shifted towards social and economic policy. Research actors have predominantly had to face indifference to their needs on the part of policymakers. In consequence, finances available for research have been cut back stringently so that the whole research system has had to suffer severe losses. To compensate these losses at least partly, many researchers and institutes have shifted their work towards contract research for industry and, what is more, away from research into other, more lucrative activities. Such urgencies clearly conflict with the aim of gaining more autonomy for scientific research. When survival is at stake, autonomy appears to be a luxury one cannot afford.

---

1 For an overview, see Uwe Schimank, Transformation of Research Systems in Central and Eastern Europe: A Coincidence of Opportunities and Trouble; in: Social Studies of Science 25/1995, 633-653.

The purpose of this book is to study these processes with regard to one particular subsector of the research system, the national academies of science. The socialist research systems consisted roughly of three subsectors. Besides the academies, there were the universities and the branch institutes. The former were mainly devoted to teaching, with research as a clearly secondary activity; and the latter were mainly conducting applied finalized research for industry, the military, agriculture, the health care system, and other extra-scientific users. In contrast, the academies were the central place for basic research[2] where large parts of the scientific elites of these countries were concentrated. As a consequence of these two characteristics, the academies were the most prestigious institutions within these countries' research systems.

This situation of the academies changed drastically after the breakdown of the socialist regime. Without going into details – which are the subject of this book – it is safe to say that, on the one hand, academy science has attained a new degree and type of autonomy on several levels. The academies as corporate actors have become more autonomous from the state; each institute within the academy has become more autonomous from the academy leadership; and each individual researcher within an academy institute has become more autonomous from the institute directors. On the other hand, the academies have also suffered from the above-mentioned resource scarcity and its consequences. Moreover, in most of these countries the academies were also confronted with a massive legitimation crisis. Having been the favorites of socialist research policy they now were the prime target of criticism within the research system. Thus, academies suddenly had to present reasons why they should continue to exist. The fact that the East German Academy of Sciences was indeed abolished very quickly aggravated this threat even more.

This book is based on empirical work conducted not just by the authors of its chapters but by a larger international research network. In the spring of 1993 a meeting was held in Bielefeld where the changes in the research systems of the former socialist countries in Central and Eastern Europe were discussed. This meeting was prepared by the editors of this book. They had been asked by Hansvolker Ziegler from the German federal ministry of research to explore the chances of building up a network of scholars from these countries who were interested in doing empirical research on the transformation of their research systems. After the meeting it was agreed that such an effort was worthwhile. The editors of this volume became the German coordinators of the merging network. They were assisted throughout by Christine Teichmann. The German federal ministry of research promoted the network by grants. The network itself consisted of individuals, often with small research teams, from twelve countries:

2 Which did not preserve them from political pressures for more applied research especially in the eigthies.

- Levan Mindeli and Elena Mirskaya from Russia,
- Gennady Nesvetailov from Belarus,
- Boris Malitzky from the Ukraine, cooperating with Alexander Nadiraschwili from Germany,
- Jan Laas from Estonia,
- Ina Dagyte from Lithuania,
- Janis Kristapson from Latvia,
- Julita Jablecka from Poland,
- Petr Machleidt from the Czech Republic,
- Eduard Sarmir from Slovakia,
- György Darvas from Hungary,
- George Zaman from Romania, and
- Kostadinka Simeonova from Bulgaria.

This network started its joint work at the end of 1993. The first phase, which lasted until autumn 1994, produced "macro" descriptions of the transformations of Central and Eastern European research systems according to a common analytical framework.[3] The second phase followed immediately. It was precisely these "macro" descriptions which stimulated our further interest. Knowing from the first phase how the finances, the personnel, the legal and organizational structures of the academy sector in comparison to the university and the branch sector changed during the first years of transformation, we now intended to understand more fully the dynamic processes of change. The findings of these investigations are reported in this book. Focussing in the second phase on the level of individual research institutes of the national academies, our general research question can be formulated like this: What has happened to these institutes as a consequences of the transformation of these countries' research systems, themselves embedded in economic and political transformations?

This question can be answered in detail only by case studies of selected institutes. The time and energy needed to carry out case studies delimits their number strongly. We decided to analyze not more than three cases in each of these countries. With an additional case from Russia, our empirical material amounts to 37 cases altogether (see the Appendix for a brief listing and description of the cases). Although this is a considerable number, it must be kept in

---

3 See the national reports in Renate Mayntz/ Uwe Schimank/ Peter Weingart (eds.), Transformation mittel- und osteuropäischer Wissenschaftssysteme. Länderberichte. Opladen 1995: Leske + Budrich.

mind that our results will not be representative. Case studies do not aim at representativity. The principal contribution of our case studies to a better understanding of the transformation processes of the research system consists of something different but equally important. By our case studies we will attain an in-depth knowledge about important "meso" processes which constitute the transformation as a "macro" phenomenon.

Although we did not want to attain representativity, the selection of cases tried to capture the variety of types of research found at the academies before transformation. Accordingly, our cases include institutes oriented towards curiosity-driven basic research with little or no immediate concern for applications, institutes oriented towards technologically relevant research which innovates the products or production processes of industry, and institutes oriented towards other, non-economic societal users of research results, for instance the health care system, or the state. To restrict the disciplinary variance, we chose in each country a physics institute, an institute from informatics, electronics, or chemistry, and either an institute of biomedicine or an institute of economics or the humanities.

For the case studies, the participating researchers used three methods: expert interviews, analysis of documents and statistics, and in some cases also a questionnaire. About five experts per institute were interviewed as a rule: present or former directors of institutes, administrative directors, and leading scientists affiliated to the institute for a long time. The interviews were conducted in the first half of 1995.

For the expert interviews and the analysis of documents we developed a common analytical framework consisting of questions to be answered for each case study. The main categories of this framework are described below. They can be divided into two groups. The first group deals with internal, the second with external aspects of the institute.

1. Internal Aspects

1.1 ACTIVITY SPECTRUM OF THE INSTITUTE

- Have the research topics of the institute changed substantially over the last few years?
- If so, were these changes self-determined, or enforced by environmental changes or actors outside of the institute?
- Which kinds of considerations were important for the changes: trends in international science, social relevance of research programs, demands of clients, especially industry? Were the changes, all in all, a shift towards "pure" or towards "mission-oriented" basic research?

- What are the decision procedures within the institute for determining research topics?
- Did the institute include other activities besides research into its activity spectrum (e.g., teaching, consulting, testing, etc.)? For what reasons?

1.2 PERSONNEL

- How has the number of researchers and non-researchers changed at the institute?
- If there have been reductions, have they been different for researchers/non-researchers, young/old, women/men? And how did the reductions happen: through management decisions (according to which criteria), through individual exit decisions?
- What is the age structure of the personnel now, in comparison to 1990?
- How has the personnel structure changed in terms of professional qualification (as measured by formal degrees)?
- To what extent do researchers on a part-time contract work in the institute?

1.3 FINANCES AND INFRASTRUCTURE

- How have the finances and the infrastructure (equipment, buildings, library etc.) of the institute changed?
- What amount of the finances comes from institutional financing, grants and research contracts? What are the sources of grants, and who are the partners for contracts? To what extent is the institute able to cover its research costs from grants and contracts?
- How is the security of finances estimated for the next three years?
- How has the personal income of researchers changed?

1.4 ORGANIZATIONAL STRUCTURES

- How has the organizational structure of the institute changed with regard to the differentiation of departments, research groups, etc.?
- How have the internal decision-making procedures (competencies of directors, participation of researchers or non-researchers) changed with regard to research topics, the composition of research teams, the allocation of finances, the income of employees? Are there significant differences between the formal procedures and their actual handling?

- Has the autonomy of researchers and research groups increased along with an increase in grants and contracts?
- What are recurring conflicts of interests within the institute? Are there structural cleavages between certain groups?

2. External aspects

2.1 RELATIONS WITH THE ACADEMY AND WITH POLITICAL ACTORS

- How have relations with political actors changed, both for the institute and for the academy as a whole?
- How are decisions of the institute embedded within plans and decisions of the academy as a whole and/or of political actors?
- How have the relations of the institute with other institutes of the academy changed?
- Are there attempts on the part of political parties, interest groups (such as trade unions) or the parliament to influence the work of the institute? If so, by which means, in which directions, and with how much success?
- How is the ministry of research intervening in the work of the institute?

2.2 RELATIONS WITH THE UNIVERSITIES

- How has the division of labor between the institute and universities changed with regard to teaching in general, the teaching of young researchers, and research? Does the institute engage in teaching?
- How many researchers of the institute engage in teaching at universities?
- Are there joint research programs, joint grants, joint utilization of laboratories or joint research groups of the institute with universities?
- Do members of the institute belong to scientific councils of universities, or do university professors belong to the scientific council of the institute?
- How many researchers of the institute have moved to universities during transformation? Have these been top researchers? Have whole research teams moved? What were the reasons?
- How many researchers from universities have, for what reasons, changed to the institute?

2.3 RELATIONS WITH INSTITUTES OF APPLIED RESEARCH

- Have the relations of the institute with institutes of applied research (especially branch institutes) changed? Are there direct cooperations?
- Are there commercial research institutes founded by members of the institutes? If so, do they use the institute's equipment or personnel? How is this regulated, and are there frequent conflicts of interest about this?

2.4 RELATIONS WITH USERS OF RESEARCH RESULTS

- Who are the actual or potential users of the research results of the institutes? Have they changed during transformation? Has their demand changed in intensity or direction?
- To what extent is the institute engaged in contract research? For whom? Who is acquiring the contracts: individual researchers, research teams, or the directors of the institute?
- Are there special transfer offices (or similar facilities) within the academy which support contacts to users? If so, is this relevant for the institute?
- How strongly, and in which ways, do political actors like the ministry of research press the institute to consider the demands of users?
- To what extent does a conflict of interests exist within the institute between the necessity to earn money through contract research, on the one hand, and the aspiration to conduct autonomous basic-oriented research, on the other? If this conflict exists, how is it resolved?

2.5 RELATIONS WITH THE SCIENTIFIC COMMUNITY, ESPECIALLY INTERNATIONALLY

- How have the opportunities to publish and to participate in conferences changed?
- Has the involvement of the institute in international research activities changed? If so, in which respects: publications abroad, participation of researchers in international conferences or in international research networks, international grants, visits from foreign researchers at the institute?
- How many researchers of the institute are currently working abroad? On a short- or long-term stay?
- If the institute acquires international grants: from whom does it do so and to what amount? Is this the result of individual initiative (requiring permis-

sion of the institute director?), the activity of the institute leaders or governmental actors, or the initiative of research partners abroad?

For each of these questions, the situation before the transformation should be compared to what has happened since. Beyond the description of possible changes we tried to identify causal factors both outside and inside the institute. What has happened because of political interventions or economic processes, as examples of external factors, and what has been the result of strategic decisions taken by the leaders of the institute, or of the typical reactions of many individual researchers, as examples of internal factors? Of course, factors of both kinds are intertwined.

The case study reports were not intended for publication;[4] we rather conceived of them as raw material for a more systematic analysis that this book is trying to offer. The case studies are all structured according to the analytical framework; but because of various kinds of difficulties in gathering the information needed not all of the questions are answered in all of the case studies, nor are the answers similarly detailed in all the cases. This limits the possibilities of comparison. Nevertheless, both striking similarities and differences have emerged; and these are the findings we want to present in this book.

The outline of the book follows roughly the analytical outline of the empirical work. It starts with *Renate Mayntz'* chapter, which elaborates the observation that the national academies and their institutes find themselves today in a political environment to which they are no longer adapted. The relevant environmental changes for the academy and academy institutes are analyzed, the challenge they have posed for their survival and the different reactions to this challenge. The chapter also asks how actors on different levels – the academy as a whole, its institutes, research groups, and individual researchers – have tried to cope with these changes. Finally the question is raised whether the concentration of research in national academies could be an efficient mode of research organization even under the changed conditions.

Julita Jablecka deals with the financial aspect of transformation. It has consisted of an overlapping of financial reductions and attempts to reallocate finances, mainly according to performance criteria of quality and/or societal relevance. This has also included changes in the modes of financing. Obviously, reduction makes reallocation all the more difficult because an institute's resistance to losses through reallocation is the stronger the more one has already suffered and expects to suffer in future from general reduction. How the interplay between internal and external factors shaped these processes, and with what

4 A collection containing one case study from each country will be published in Russian. The volume will be edited by Levan Mindeli and Alexander Nadiraschwili.

consequence for salaries, infrastructure and other research conditions, is the topic of this chapter.

Stanislaw Provaznik, Petr Machleidt and Adolf Filácek study how the reduction of personnel that took place in almost all academy institutes has come about. They describe different aspects of reduction and detect factors which have determined its magnitude in the different cases. The reductions have been handled very differently. Roughly, two main strategies can be distinguished. In many cases there have been almost no outright dismissals. Instead, measures such as collective reduction of work-time, reduction in salaries and long-term unpaid vacations have been taken. Only individuals made exit-decisions if they perceived better job opportunities for themselves outside the institute. As an alternative, in some countries there have been attempts at a more active adaptation of the institutes' size and structure of personnel. In a deliberate search for an increase in the institutes' competitiveness, personnel has been dismissed on the basis of performance evaluations.

György Darvas considers in his chapter how the institutes changed their type of research during transformation. He detects a common path, winding back and forth between applied and basic research. Many institutes took the opportunity of the breakdown of the socialist regime to turn from applied research, which they had been pressed to do before, to basic research. But financial difficulties motivated them to turn back to applied research again, in search of extra-scientific users of their research results who would be willing and able to pay for them. When the institutes realized that there was not much additional money to earn in this way they returned to basic research again. Thus, most institutes have not yet reached a stable equilibrium of internal preferences and external pressures and opportunities with regard to their type of research.

The next chapter by *Uwe Schimank* inquires about the major determinants of an institute's fate in the transformation dynamics. More specifically, it reflects upon the surprising fact that three factors apparently have no strong causal influence on an institute's fate in transformation: Whether an academy institute does more basic-oriented or more applied research, whether it is a natural science or a social science institute, and whether it does high or low quality research. In many cases these distinctions do not greatly affect an institute's relative success or failure. Several additional factors are drawn from the case studies for a more adequate explanation of an institute's fate.

Eduard Sarmir and Stefan Zajac describe how the relations between the academies and the universities have developed during transformation. The collapse of socialist rule at first reversed this relation, the academies losing prestige and having to defend themselves, with the universities demanding and gaining a new status. But after a while the initial turbulences calmed down in most of the countries, and not much has changed between universities and academies. Universities have had to accept that there is no chance of eliminating the academies and incorporating their research capacities. Remarkably, though, contacts and cooperation on an informal, personal level seem to have increased.

Gennady Nesvetailov deals with the fact that in the reduction of research personnel, young researchers especially have left because they have either had the best outside opportunities or been easiest to dismiss, and it has not been possible to recruit any new ones. Moreover, research is not an attractive career anymore. This is not only a consequence of unsatisfactory career prospects, but also of a serious loss of social prestige on the part of the research profession. Therefore, academy institutes face the very serious problem of the collective aging of their personnel. This chapter, firstly, describes the extent of this aging and its dysfunctional consequences for research. Secondly, it asks which strategies are adopted by some institutes to work against this aging.

Elena Mirskaya's topic is the changes in the international relations of researchers and institutes. While different forms of communication and cooperation with Western partners have intensified, at the same time the established traditional international networks among socialist countries have seriously decayed almost everywhere because of inadequate resources. In addition, it can be shown, at least for Russia, that a new form of stratification of scientific communities has taken place in which a researcher's intensity of contacts with Western partners plays an important role. This line of stratification cuts across the formal hierarchies and professional career paths.

The next chapter probes into the relationship between an academy institute and its academy. *Kostadinka Simeonova* argues that there is now an ambivalent view of the academy on the part of the institutes. On the one hand, the academy is still often seen as bureaucratic and rigid and in need of reform. The institutes, however, are no longer affected by this inflexibility to any great extent because their autonomy has grown, especially with respect to their respective research programs. On the other hand, the academy is seen by the institutes as a necessary protective shield against external pressure, and as a device of collective self-protection. Thus, it seems that in many cases a new modus vivendi between institutes and academy has been established.

Finally, *Peter Weingart,* taking case studies from Russia and the Czech Republic as examples, sketches a striking difference of mood, in which academy institutes from different countries cope with the challenges of transformation. One may differentiate between "offensive innovative" and "defensive conservative" strategies of transformation, as well as between self-critical and self-deceptive evaluations. It appears that a combination of self-critical self-conception and more radical strategies of transformation account for a more optimistic outlook, which may well be an important condition for ultimate success.

It would make little sense for us to summarize the results of all these chapters in some kind of simple pattern. Each chapter draws a rather complex picture of its respective aspect. We think that this is one of the strengths of the work presented here. There simply is no easy way to understand what happens in these countries' transformation dynamics in general, and in the transformations of the academies in particular.

Köln, Hagen, Bielefeld, March 1998

Renate Mayntz
Uwe Schimank
Peter Weingart

THE IMPACT OF RADICAL REGIME CHANGE ON THE EAST EUROPEAN ACADEMIES OF SCIENCES

RENATE MAYNTZ

Max Planck Institute
for the Study of Societies
D-50677 Köln

The organization of research in national academies of science is a Soviet invention. It was developed in Soviet Russia and then extended to other Soviet Republics, including the three Baltic states after they had been forcibly integrated into the Soviet Union. The various regional academies in the Soviet Union formed an interrelated and in part functionally differentiated system that was directed by the Soviet Academy of Sciences in Moscow (the Russian Soviet Republic did then not have a separate academy). After the Second World War all East and Middle European satellite countries sooner or later copied the Soviet model of research organization. Most of these countries used to have academies in the form of learned societies, but their universities were the major sites of basic research. To that extent they stood in the West European tradition. The introduction of the Soviet model of science organization, where academies served as the major institutions of state financed research, broke with that tradition. In the highly developed Western industrial countries, academies typically continued to exist as learned societies, while state-supported non-university research organizations grew up outside of them. Thus in Germany the Kaiser-Wilhelm-Gesellschaft, founded in 1911 and continued after the Second World War as Max-Planck-Society, has become the largest German organization doing basic research. Even the English Royal Society which played a significant role in the development of the experimental natural sciences in the 17th and 18th centuries did not grow into a large research organization; instead, the promotion of basic and problem-oriented basic research became the task of the various British Research Councils.

There is an evident fit between the Soviet type of research organization and the state socialist society. Centralized state control, division of labor, and the concentration of similar tasks in one large organization have been the principles underlying the structure of socialist economies. The concentration of basic and problem-oriented research of most or even all disciplines in one large organization is fully congruent with these principles. Also, the Soviet type academy is

R. Mayntz et al. (eds.), East European Academies in Transition, 1–12.

directly subject to state control and thus conforms to the hierarchical governance mode, while the traditional learned societies were based on the principle of scientific self-organization.

But more than a congruence of organizational principles is involved when we speak of a basic fit between state socialism and the academy model of research organization. In "scientific" socialism, science was considered the basis of production and planned innovation. Its key function in the process of societal reproduction and development gave science a special prestige and justified both an elite status of scientists and special state support to scientific research. Provaznik et al. (in this volume) observe that this general view of the reproduction process as scientifically based justified a high share of R&D personnel among the gainfully employed and thus explains the overexpansion of R&D staff in socialist as compared to Western capitalist countries. Along the same line, OECD experts remark that the Russian S&T base was extremely large by comparative international standards, whether in terms of employment, number of establishments, or resource commitment as a share of GNP, and that this expansion of S&T institutions was driven by political rather than economic forces (OECD 1994: 15-16).

As Provaznik et al. (this volume) correctly note, the policy guideline „the development of science is to be secured in advance of production“ corresponds to a linear model of technological development. The linear model of technological development links basic and applied research and the practical application of research results in one causal chain. This model came to supplant the earlier image of a vertical rank order relationship between theory (philosophy) and practice in the course of the 19th century, parallel to the gradual shift from a stratified to a functionally differentiated society (Mayntz 1997). Functional differentiation means division of labor on the macro-level of a society, i.e. the existence of separate institutional complexes for different purposes such as education, health, production of goods and services, etc. There are, however, different modes of relating the various functional subsystems to each other. In Western democratic market societies the functional subsystems are supposed to adapt spontaneously to each other, following rules of exchange; in state socialist societies, in contrast, the functional subsystems are coordinated by a central authority. In both types of functionally differentiated society, the science subsystem is seen to produce important inputs for other subsystems such as the economy, the health system, technical infrastructure systems, etc. But while this is perceived as a spontaneous exchange relationship in the Western type of society, the production of knowledge is planned and centrally controlled in state socialism. Accordingly, we find two different versions of the linear model of technological development.

In the Western type of society, basic research, the first link in the chain, is the driving force of the process of technological development; this version of the linear model is fittingly called "science push model". It is based on the assumption, formulated maybe most clearly by President Roosevelt's science advisor Vannevar Bush at the end of World War II, that scientific research is most creative and productive when it is left to follow its own dynamic (Bush 1945). This implicit theory largely guided Western science policy during the first two postwar decades. In contrast, the view that scientific research must be politically directed toward desirable ends goes hand in hand with a „demand pull model", in which desired applications drive the search for knowledge. This version of the linear model lay at the base of science policy in socialist countries, where research was centrally planned and organized as part of the societal production and reproduction process. The academy model, i.e. the concentration of research in one large, specialized, and politically controlled organization[1] is thus closely linked to a particular view of the function of science in a planned society and the proper relationship between knowledge production and application. In this view of science and society, there is no place for that autonomy of research so dear to the heart of scientists and politicians in Western, democratic market societies.

The chain reaction of reforms and – mostly rather peaceful – revolutions that swept Central and Eastern Europe towards the end of the 1980s did not only overthrow the state socialist regimes; it also undermined the basic logic of the academy model of research organization. Three related changes in the environment of the science system should be pointed out in this connection. The first was the attempt to emulate the economically more successful Western democracies by abolishing the communist party's monopoly and claim to comprehensive central control, and by introducing market principles in the economy. Where production is no longer politically controlled and market principles come to reign, the justification for claims of centralized research planning is lost. Where different functional subsystems of society are granted a certain autonomy, the science system, too, is emancipated from central control.

A second, related change has taken place at the level of political priorities. This change followed from the difficult economic situation in which all transformation societies find themselves and which has given priority to economic survival and the avoidance of social unrest, leaving little room for future-oriented investments. The economic problems besetting the transformation societies are such that science can do little for their solution. So far, and in contrast to capitalist countries with whom they must now compete on the world market, the former socialist countries try to solve their economic problems, all programmatic statements notwithstanding, by offering cheap labor for mass

---

1 It does not contradict the validity of this statement to observe that the scientific establishment in socialist societies was actively involved in the process of research planning. Political control was not a simple top-down process, but contained elements of bottom-up processes of planning within the politically defined framework of binding values and goals. A detailed description of this control structure for East Germany can be found in Kocka (1997).

production rather than by high-tech production and innovation. At the same time, the newly opened frontiers obviate the former need to reinvent Western technological achievements that could not be imported because of the Western embargo or due to the shortage of hard currencies. Both developments have lowered the demand for R&D and detract from the practical value of science. Science, therefore, does not receive a privileged treatment anymore. In a situation of general scarcity science budgets have been cut overproportionately, and issues of institutional reform are treated slowly and sometimes erratically.

The third factor is a parallel development at the level of values. The high prestige enjoyed by science as long as it was considered the functional basis of progress and the prime condition of reaching and even overtaking the capitalist countries of the West has decreased with the erosion of the belief in all forms of central planning. The shift towards more individualistic values of personal well-being has reinforced this decrease in the value of knowledge.

The effects of these changes in the political, economic, and cultural context on the science system are described in detail in the different chapters of this book. Though it is not possible to separate fully the consequences of economic, institutional, and value changes from each other, the case studies on which this book is based make clear that they have contradictory effects: While the legal and institutional changes have generally increased the independence of science and have thus enlarged the scope for action enjoyed by scientific institutions, the economic situation and the attendant financial difficulties have seriously restricted this scope. Thus, the right to select research topics independently is of little practical value where institutes and scientists are forced, for financial reasons, to concentrate on work for which they can get a contract or grant, and the new freedom to travel comes to naught where there is no money to pay for it. The changes we observe, therefore, are only partly reforms, and in many respects merely adaptations to a troublesome situation. In the following section, however, we shall concentrate on changes that fall into the category of „reforms“ – deliberate and goal-oriented institutional changes in the science system, especially as they touch upon the role of academies. The most important of these changes can be briefly summarized in six points (see especially the chapters by Jablecka, Provacnik et al., Sarmir, and Simeonova). It should be noted, however, that these points are only rough indications of reform *directions*, and that the corresponding reforms have not been realized in all of the countries included in this study, or have been realized in different ways and to a different extent.

(1) Central control and political guidance of the science system have greatly diminished, even if not entirely disappeared. This is in large part an effect of the political regime change – the loss in directive capacity of the former ruling party, the introduction of a multi-party system, and the limitations in government powers resulting from a democratic constitution. But scarce financial resources and shifting political priorities have also played a role.

Often, the independent status that science has gained is even written into law. An important consequence is the academy's official freedom to define its own research program.

(2) In many ex-socialist countries, the national academy of sciences has lost its privileged position within the science system, both with respect to its science policy functions and with respect to its position vis-à-vis the universities. Formerly, the academies were heavily involved in developing the national research plans. Now, the locus of science policy making has shifted to government agencies; in several cases, ministries responsible for science policy and the promotion of research have newly been formed. This shift, however, does not mean a radical change since academy scientists, who not only enjoy scientific reputation but dispose of valuable experience in research management, still tend to play an important role as expert advisors. In some countries they have been and are actively involved in restructuring the science system and introducing the various reforms. Nevertheless, academy institutes can no longer claim preferential treatment, but have become one of several sectors in the science system commanding equal attention. Universities in particular, who used to feel underprivileged as research sites, have been able to improve their status and have sometimes been less afflicted by personnel reduction than academy institutes. However, the cooperative relations of academy institutes with universities, which often did exist already before, do not seem to have become much more intensive. At times it seems that there is even more competition now, especially where both types of institution look to the same sources for research grants.

(3) The academy leadership has lost much of its programmatic steering function for the academy institutes, and within limits, the same is true for the relationship between institute leaders and individual scientists or research teams. Thus, the process of decentralization observed under point (1) at the national level continues at the intraorganizational level. This process is in part deliberately planned, and finds expression for instance in the establishment of new participatory (democratic) procedures and bodies. In part, however, it is an unintended side effect of the weakening of the leadership's resources and in particular of the scarcity of centrally available funds, which forces institutes and individual scientists to engage in fund raising activities for themselves by themselves.

(4) Where the academy before the regime change served both as a research organization and a learned society, these two components tend to become more distinctly separate. Thus the institutes may have a representative body of their own at the level of the academy, which now fulfills coordinative and strategic guidance functions for them that pertained to the academy leadership before. In the case of the Czech Republic, the learned society was even dissolved as part of the academy in 1992, and re-established

a year later as a private association, while the academy continues as a research organization.[2]

(5) Competition for resources, not only between the different sectors of the science system but also between different institutes and research teams, has been introduced as a measure to increase the overall quality of research. The institutional form in which this takes place is typically the establishment of national. In several countries, these funds dispose of a growing share of the total R&D budget. Expert scientists play a crucial role in these funds. In principle, this mode of R&D financing increases the autonomy of the science system.

(6) At least in the first phase of the reforms, the academies have programmatically emphasized their basic research orientation. This reorientation is at least in part a reaction to the strong pressure towards applied research that characterized socialist science policy especially during the 80ies. In some countries, notably Romania, the imperative demand that science must be directly useful even became a threat to the very existence of the academy, which lost control over its institutes in 1974 and could not even elect new members to the learned society, being considered recalcitrant to demands for full political control over, and practical usefulness of all scientific research. After the regime change, Romanian research institutes were in contrast required to perform mainly basic research to qualify for membership in the revived academy.[3] A basic research orientation can contribute to the autonomy of science, because it diminishes its dependence on specific user groups in defining research goals. At the level of individual institutes, the basic research orientation was reinforced by evaluation procedures giving priority to theoretical importance and cognitive innovativeness of research, and by the chance to acquire grants from (possibly foreign) funds favoring basic over applied research. On the other hand, the need to find additional income from contract research has sometimes induced institutes to engage in more applied research than they would have done otherwise.[4] In some countries where former communist forces have recently regained power, science policy appears to shift again generally to a stronger emphasis on the practical applicability of science.

The reorganization of the science system along these lines did not necessarily imply to dissolve the existing academies as research organizations. This happened, in fact, only in East Germany and in the three Baltic Republics, but it

---

2 Information in a letter by A. Filácek and P. Machleidt.

3 See the (unpublished) case study report by Gheorghe Zaman, Steliana Sandu, Anca Dacin: Transformation der Rumänischen Akademie. Bukarest 1995.

4 This pressure was, of course, much stronger in the case of formally privatized research institutions, especially formerly state financed branch institutes that used to serve industry; see Couderc (1996) for details on Russia.

was also debated in other countries, especially in the Czech and Slovak Republics, in Hungary and in Poland. University representatives tended to be the most outspoken critics of the academies' dominant role in research, but there was a more widespread criticism of the central control, the bureaucratic rigidity, and the inefficiency which were attributed to the existing academies of science. To take account of these specific criticisms, however, did not require the dissolution of the academies, since significant improvements could already be expected from the reforms summarized above. In fact, in the four countries that rejected the academy model completely there were additional reasons for doing so.

In East Germany, the former Academy of the Sciences has not even survived as learned society, and only one third of its institutes have preserved their institutional identity more or less intact as they have passed into the hands of new financial carriers (Mayntz 1994: 195). In the three Baltic Republics, the academies have been divested of their institutes, which have typically become subject to the jurisdiction of a ministry, but they continue as learned societies and, as in the case of Estonia, still serve as spokesman and interest representative for their former institutes. These particularly radical transformations were the result of specific factors not present in any of the other countries. In the German case, this factor lay in the special circumstance of transformation by integration into another state. Legally defined as accession, this integration meant that East German institutions must be adapted to West German structures. As major organization of non-university research, the East German academy of sciences was incompatible with these structures. In the case of the three Baltic Republics, it was their political history that provided the motive for a radical change away from the Soviet model. These three countries had been incorporated against their will into the Soviet Union, and a Soviet type academy had subsequently been imposed upon them. It was thus part of the act of national liberation that a new model of research organization was chosen.

The general direction of the institutional changes in the science system of the former socialist societies is in accordance with the overall change in societal governance they are undergoing. Subjectively, however, these changes were experienced as a radical modification, if not a complete breakdown of the familiar order. Such situations are ripe with both opportunities and threats. In the beginning, the opportunities may have appeared more important, notably the opportunity to gain more autonomy for universities, academies, and academy institutes. Universities also used the opportunity to challenge the dominant position of the academy. Soon, however, financial troubles became more prominent than opportunities for institutional change; the socio-economic environment of science became turbulent and threatening. The sources of these troubles were twofold and hit the different sectors of the science system with unequal severity: the cessation or at least big cuts in state funding (at least in terms of real value), and a steep decrease in the money available for contract research, especially on the part of industry. The consequences were shortage or even complete lack of money for investments in the research infrastructure, the need to reduce personnel, the inability to pay decent salaries to the research staff, and the sharply decreasing attractiveness of a career in science both for existing qualified researchers and for new recruits. Research establishments that were privatized clearly suffered the greatest shock, but even where state funding continued as in the case of the academies, the economic transformation process had dire consequences for scientific research that are amply documented in the chapters of this volume.

To cope with the crisis situation called for new strategies, but the chances for successful coping varied between academies and institutes as corporate actors and researchers as individual actors.[5] Demonstrated conformity with political wishes brought no longer substantive rewards, nor did skilful ideological impression management do the job. Since the trouble that now afflicted them had very little to do with the previous performance of the academies, their willingness to accept and even voluntarily initiate reforms was not able to neutralize the basic threat in the form of financial shortage. Nor was scientific excellence a certain safeguard for research institutes (see Schimank in this volume). Coalition building with state actors could help to alleviate the crisis for academies, but this strategy was available only where there was a political bond between the reform forces in politics and in science, which was not true in all countries. Besides, even where such coalition building was feasible it could not off-set the negative effects of the economic crisis. The political support game also became more difficult to play for the representatives of science because political power now tended to be more diffuse. In general, science now lacks a strong political anchor and a powerful lobby.

The academies as corporate actors thus could do little to ameliorate their situation by attempting to influence their environment; they had to seek ways to adapt by trying to minimize the damage. One strategy of adaptive coping was rationalization to save money and increase productivity; this strategy was chosen for instance by the Czech academy where personnel reduction was not passively suffered, but actively embraced as a means to increase effectiveness. Along the same lines is the selective operation of budget cuts following the evaluation of institutes. A different strategy, typical especially for the Russian academy, is followed where academies allot budgets to the institutes on the basis of their (nominal) personnel strength, thus impeding the making of personnel cuts and leaving the institutes to their own devices in trying to survive on the basis of manifestly insufficient resources. The impotence of the academy as corporate actor to assure the survival, let alone the well-being of the institutes leads to a de facto decentralization, reducing the internal control capacity of the academy as a consequence of its loss of control over its environment. This is true even where an academy attempts a more active strategy, as in Hungary,

5 See Schimank/ Stucke 1994 for a general discussion of coping with trouble in science.

where „the institutes were compelled to find their own ways to identify their 'customers'", since the academy lacks an organ to fulfill this function for them.[6]

In the transformation process, the survival of individual institutes is not fully tied to the survival of the academy as an organization. In principle, the academy may be dissolved as such but some or even all of its institutes may well survive. On the other hand, the persistence of the academy does not necessarily assure the survival of all of its institutes. As the case studies which form the data base of this book show, the political and economic changes did not affect all academy institutes in a given country in the same way,[7] and there is a great variety in the amount of resource cuts they had to cope with. But by and large, it was the same bundle of coping strategies that could lead to a (relatively) successful adaptation to a threatening situation. (1) One strategy was to change the institute's research profile in accordance with changes in external demand and criteria of relevance. (2) A second strategy was to reduce personnel selectively, so as to keep a maximum of the productive researchers and create openings for young scientists. A repeated evaluation of scientists helped in this effort. (3) Finally, institutes could actively look for „new customers" and for new sources of incomes. At least the first and the third of these strategies can, however, have disfunctional consequences for the quality of research. This is true for instance if additional income is generated by renting part of the institute's premises or by engaging in commercial activities, or if in the search for new customers the institute's own research program is compromised. Though some institutes for instance in Poland have tried it, there is also little scientific reward in turning to teaching and training. Strategies assuring the short-term survival of an institute may thus prove disfunctional for the quality of science in the long run.

Somewhat surprisingly, the case studies that form the data base of this book show that the weakest actor of all, the individual scientist, may have the largest scope for coping. In contrast to the larger social units, individuals have, above all, an exit option: they may look for another job in their own or in a foreign country, in science or in some other field. Individual scientists also have the opportunity to find funds from other sources:national and international grants are normally negotiated, applied for, and given to individual researchers or small teams rather than to an institute. For the institute as a corporate actor, the successful acquisition of grants and contracts by its members is, however, a double-edged sword. On the one hand, this is an important source of income and permits to supplement the scarce salaries. On the other hand, this form of successful coping implies a de facto decentralization to the level of small groups

6 György Darvas: Consequences of the Transformation of Central and East European Science Systems for Three Institutes of the Hungarian Academy of Sciences. Budapest 1995, 9.

7 For a systematic analysis of the varied fates of East German academy institutes see Wolf (1995).

or teams within the institute, and a consequent fragmentation of its program structure and weakening of the leadership's directive capacity with respect to the choice of research topics. What we here observe is more than a special feature of the particular transformation process under review; the erosion of larger, hierarchical structures is exactly what typically happens in highly threatening and turbulent situations, that occur not only in revolutions but also after being defeated in war.

Successful coping may secure survival, but its aggregate effect need not be positive for the structure and functioning of the emerging science system. Thus individual success in finding employment in foreign research institutions constitutes the negative phenomenon of „brain drain" at the system level. Undoubtedly some of the measures aiming to increase the efficiency of research institutes, such as funding according to criteria of evaluation and competitive bidding for research grants, will have a positive effect. The same holds for the improved conditions of contact and exchange with the international science community. These positive effects, however, may easily be offset by the deterioration of the research infrastructure due to the lack of investments, and by the aging of the R&D staff described and analysed in detail by Nesvetailov in this book.

A question of particular importance in the context of this book is whether or not the formal dissolution of academies as „holding organizations" for state-financed research institutes is to be judged positively, so that their continued existence would be a sign of half-hearted and incomplete reform, as for instance Karel Müller (1995: 53) seems to believe. In fact there are experts in the countries that have maintained their academies as research organizations who would favor their dissolution, and OECD experts have even explicitly recommended the separation of the scientific establishments from the learned society of the academies of the Czech and Slovak Republics (OECD 1992: 67). The argument for such a separation that OECD experts have spelled out in another report concerning Hungary revolves basically about „role overload". Arguing that the Hungarian Academy of Sciences should „choose among the many different roles that it used, and to a certain extent, continues to play", the experts list five different functions: giving expert advice to government, representing the interests of the scientific community, giving accreditation to doctoral degrees, providing money for basic research, and running research institutes. „The concentration of several of these different functions in a single organism is not safe", the experts continue. „In particular, the simultaneous presence of both the advisory/ lobbying function and the funding/management function presents a potential conflict: the former requires independence while the latter requires some dependence on the government and the parliament to which the body, and more precisely its managers, should be accountable for the money spent on it." (OECD 1993:112-113) While this conflict is mitigated where academies loose their former privileged position within the science system and research financing is shifted to national funds, there remains a potential conflict with the universities over the accreditation of degrees. Also, there exists a structural tension between the academy as a research organization and the academy as a learned society. This is indicated by the fact that in the course of academy reforms, a previously existing hierarchical relationship between the body of academy members and the research institutes has typically been abolished, so that learned society and research institutes, if they still co-exist in one organization, are now more loosely connected. If separation is the spontaneous tendency, one may well question the functionality of maintaining both under the same institutional roof.

But even if the argument for more functional differentation seems convincing, this does not mean that the solution found in the Baltic Republics, i.e. the continuation of the academies as learned societies and consultative organs, and the transfer of their formally now independent institutes to ministerial jurisdiction, is optimal. In the major Western countries, the operative work of promoting basic and problem-oriented basic research is rarely performed by a ministry. Instead, *independent* publicly financed agencies are usually assigned this task. In these agencies, scientists typically play an important, even dominant role in the decision processes. There are good reasons for such an institutional solution, even if historically speaking these structures have not been developed out of rational political considerations, but are often the result of successful lobbying on the part of influential representatives of science (see Braun 1996). The functional arguments for this institutional choice refer, firstly, to the insufficient information basis of political actors for an effective guidance of science, and secondly to the fact that within the science system, the production of knowledge is regulated by the operation of (institutionalized as well as informal) peer review – a mechanism that politicians and clients of knowledge producers can make use of, but that they are unable to operate themselves. For such reasons it is doubtful that to subject former academy institutes to the control of a ministry will in the long run prove successful. In fact, such criticism can even now be heard from experts in the Baltic Republics.

It seems, then, that more functional differentiation would have to go hand in hand with granting the major organizations of research that measure of autonomy typically enjoyed by the publicly financed non-state institutions in Western countries that promote basic and problem-oriented research outside of the universities. The "reformed academy" model that we find today e.g. in Hungary and in the Czech and Slovak Republics seems to move towards such a solution, without as yet having fully achieved it.

References

Braun, Dietmar (1996). *Die forschungspolitische Steuerung der Wissenschaft*. Frankfurt/M: Campus (in preparation).

Bush, Vannevar (1945 [1960]): *Science, The Endless Frontier: A Report to the President*. Washington, D.C.: National Science Foundation.

Couderc, Marie-Laure (1996). Adaptation of some former research units to the new economic environment in Russia. *Science and Public Policy* 23, 375-384.

Kocka, Jürgen (1997). Wissenschaft und Politik in der DDR. Draft chapter for the final report of the working group *Wissenschaften und Wiedervereinigung* of the Berlin-Brandenburgische Akademie der Wissenschaften.

Mayntz, Renate (1994). *Deutsche Forschung im Einigunsprozeß: Die Transformation der Akademie der Wissenschaften der DDR 1989-1992*. Frankfurt/M.: Campus.

Mayntz, Renate (1997). Forschung als Dienstleistung? Zur gesellschaftlichen Einbettung der Wissenschaft. In: Berichte und Abhandlungen der Berlin-Brandenburgischen Akademie der Wissenschaften, Vol. 3. Berlin: Akademie Verlag, 135-154.

Müller, Karel (1995). Transformation of National Research Systems in the Countries of Central Eastern Europe. *Science Studies* 8, 44-55.

OECD (1992). *Reviews of National Science and Technology Policy: Czech and Slovak Federal Republic*. Paris: OECD.

OECD (1993). *Science, Technology and Innovation Policies. Hungary*. Paris: OECD.

OECD (1994). *Science, Technology and Innovation Policies. Federation of Russia*, Volum.e I. Paris: OECD.

Schimank, Uwe, & Andreas Stucke (eds.), (1994). *Coping with Trouble: How Science Reacts to Political Disturbances of Research Conditions*. Frankfurt/M.: Campus.

Wolf, Hans-Georg (1995). An Academy in Transition: Organizational Success and Failure in the Process of German Unification. *Social Studies of Science* 25, 829-52.

FINANCIAL THREATS TO ORGANIZATIONAL SURVIVAL

JULITA JABLECKA

University of Warsaw
Center for Science Policy
and Higher Education
PL-00-046 Warsaw

1. Introduction

The transformation of research institutes at the Academies of Sciences constitutes a dynamic element in the changing social and economic system of post-communist countries. Changes of the scope, form and structure of financial support for research activity have become an integral component of the general reforms. At the same time, a significant deterioration in the financial support available to research and educational institutes has characterised the financial picture of the post-communist countries throughout the 1990s.

The question arises as to whether the decline in funding is an unavoidable – albeit unintentional – consequence of the whole system of reforms or whether it is the result of a deliberate policy aimed at curtailing a state research sector which was seriously overdeveloped under the previous regime and which now requires pruning to fit the newly-emergent market economies of the post-communist countries. Since all the institutes examined during the course of our study have had to grapple with the problems of reduced funding, a further question must be addressed. To what extent is the relatively poor financial health of the surveyed institutes caused by changes to the system of research funding, and to what extent is it caused by the institutes' (in)ability to adapt themselves to the changed situation? What impact have the changes in the funding structure had on institutes' management of their financial affairs?

Analysing the financial circumstances of the institutes is not a simple matter. It is difficult to identify a precise moment when the era of socialist science ended, to be replaced by reforms aimed at adjusting to the requirements of a democratic polity and a market economy. In some countries, a radical approach was taken whereby the reorganisation and refinancing of research was tackled as a whole. In others, a more gradual and incremental approach has been preferred whereby improvements continue to be introduced piecemeal. In addition, we find differences across the different countries as to the depth and the precise

R. Mayntz et al. (eds.), East European Academies in Transition, 13–31.

direction the reforms to the political and economic systems are taking. The exact conditions brought about during the period of transformation, with each national system possessing its own unique context for reform, must be borne in mind when we analyse the changes being experienced by research institutes as they adjust to their new environments.

As a result of financial reforms and more widespread economic turmoil, we find varying forms of funding re-allocation between institutes; differing sources of support for institutes; and, within the institutes themselves, differing approaches to restructuring and reorganising their scientific work. In addition, we find variations in the changing relationships between the Academies and their respective institutes across the different countries.

2. Problems of Governmental Research Funding in the Period of Transformation

Our analysis of changes in the financial circumstances of research institutes covers the period 1990 to 1994/5. We differentiate between the early period, characterised by the collapse of the centrally-planned economy and the immediate impact of that change on research funding, and the following transitional period, when economic change was accompanied by more systematic reforms of research institutes.

In the majority of the Central and Eastern European countries, the immediate consequence to research institutes of the collapse of "real socialism" and the socialist economy was a sudden and shocking deterioration of financial support as compared with the monies available in the 1980s. All the various forms of state provision for research had their budgets squeezed as both the sources and the mechanisms for financial support underwent change; not only was funding reduced but the allocation mechanisms also changed at the same time. Subsequently, and in almost all cases, the financial health of research institutes deteriorated further. However, these later changes were experienced at different levels of intensity in different countries and the pace of change varied across individual institutes. In some rare instances, it was even the case that the decrease in revenues slowed down or – sometimes – improved slightly.

The general deterioration of the financial circumstances of the surveyed institutes is intimately linked with more widespread changes within the socio-economic and political environments of the relevant countries. The transition to a market economy has typically been accompanied by phenomena such as inflation, price fluctuations, decreased budget revenues, devaluation of local currencies relative to the dollar or the replacement of local currencies, and recession. As a result, even if the face value of subsidies for institutes has not altered during this period, their real value has dropped dramatically. The authors of national reports have also emphasised that shrinking state budgets have had the effect of reducing the monies available to R&D as a proportion of national GDP. All of these external sources of financial turmoil have contributed to the continuing funding difficulties experienced by research establishments.

A particularly severe example of economic crisis, and its effect on cuts to subsidies, is provided by Belarus. During the 1990s, inflation sky-rocketed such that by the end of 1994 inflation was running as high as 28% per month. By mid-1995 this runaway had been restrained to 2%/3% per month but difficulties were intensified by a budgetary deficit which amounted to 3% in 1994. (By 1995, the deficit was reduced to 1.8%.) Simultaneously, the GDP level decreased in comparable prices; between January and June 1995, it dropped by 12% on the previous year. The science sector suffered the consequences of this turmoil more than other societal sectors: R&D expenditure as a proportion of GDP decreased from 2.23% in 1991 to 1.47% in 1993 and 1.35% in 1994. A similar scenario of economic crisis emerged in the Ukraine, in Russia, and in some other countries of the ex-Soviet Union. In the case of the former Soviet republics, the collapse of the USSR marked a turning-point in the financial relationship between the Academies' institutes and their contracts with industry and with commercial businesses in the other Soviet republics on the one hand, and, on the other, with the branches of the formerly centralised USSR ministries. Changes of this kind in the macro-economic system sometimes retarded the ability of scientific institutes to retain financial control.[1]

Hungary provides us with an example which illustrates other effects of the widespread financial crises experienced by the transitional economies. During 1990 to 1994, economic reforms which had been initiated in the 1980s continued to be put in place. Nevertheless, during 1990 to 1994 – and despite the fact that research conditions in Hungary did not deteriorate in the same drastic way as had been experienced in other post-communist countries over this time – R&D expenditures decreased both in real terms (by 50%) and as an expression of their proportion of GDP (by 20% in real value).

How did the institutes of the Academies fare, as compared with other higher education and governmental bodies? In Hungary, the Academy institutes enjoyed a slightly more advantageous position than did the Academy institutes of other countries. Government-funded research at the Hungarian Academy only dropped by 31% in 1994 from the funding level available in 1990. This compared with a 50% decrease in overall state funding of R&D activity, and it was the ministerial institutes in Hungary which bore the major brunt of the funding cuts.

The Czech and Slovak Republics and Poland also provide examples of countries whose R&D activities have been less severely affected by financial restraints than other countries in the study. However, as far as the institutes of

1 For example, in Latvia, 1992 and 1993 were the worst years for research institutes in financial terms. Since 1993, in Estonia, inflation has been running at a level of 25%/30% annually.

the Academies are concerned, in the majority of countries, funding for the Academies' institutes decreased more steeply than funding for other higher educational establishments. In the Slovak Republic in 1994, monies allocated to the Academy amounted to 9% of the state's total national expenditure on R&D as compared with 14.6% in 1990. As a result, despite staff reductions of 44.8%, expenditure per employee declined by 27.8%.

In Poland, 1991 was the first year the new framework for funding research came into effect, although it was also the second year of operation of inflation restraint mechanisms. If we take the expenses of funding all research activity for 1991 as a baseline of 100, then in 1994 – taking inflation into account – these expenses amounted to 76.8% of the sum of 1991. Those for the Academy's institutes amounted to 60%, with a simultaneous decrease in employment in this research sector of 15%.

The funding sources available to the Academy institutes in the post-communist countries are not limited to state subsidies. Prior to the reforms, industrial ministries and government bodies had provided alternative sources which the Academies' institutes tapped. However, the process of depleted budgetary funding post-1990 was also accompanied by the drying-up of funds available from these alternative sources; after 1990, institutes could avail of fewer opportunities to commission orders directly from industry. It was only very rarely that such industrial contracts increased during this period. Funding from this kind of source was, in general, negligible.

The deterioration of the financial situation of the research institutes during the transformation period reflects more than just the more widespread economic difficulties experienced by the post-communist countries. Another reason proposed by some national case studies is that the general prestige of science also declined during this period or, at least, that political interest in supporting scientific effort virtually disappeared. While science had enjoyed high prestige under communist rule, its elevated status has diminished in the new post-communist environment.

3. External Factors Affecting the Transformation of R&D and Their Impact on Institutes

The budgetary constraints on the economies of Central and Eastern Europe have had quite a homogeneous impact on the deteriorating financial stability of all the Academies and their respective institutes. An important factor determining the financial circumstances of the institutes we surveyed concerned the policy each Academy pursued to regulate how its institutes received whatever monies were available, and how those funds should be allocated across and within its institutes.

In some countries, the Academy did not differentiate between its institutes and simply reduced the funding for all of them across the board. The Ukraine provides us with an example of this policy. Between 1990 and 1994, outlays on research within the Academy dropped by one-tenth. At present, funds are mainly diverted into salaries and municipal services. In common with some other countries, the guiding principle in the Ukraine Academy's policy has been to protect the intellectual elite of the nation – a principle reflected in the protection offered by the President's ordinance of 1994 whereby Academy members are permitted supplements to pay awards. However, the rampant inflation of the Ukraine led to the relative reduction of government subsidies to research institutes; reductions pertained to all institutes equally and cuts were executed homogeneously on a per-employee basis. The Ukraine institute of chemistry offers us an example of the consequences of this equalising policy. While its share of state subsidy has not nominally decreased (it remains at 50% of the institute's total revenues) its real value under conditions of high inflation has plummeted. Meanwhile, this institute cannot adequately compensate for the loss by tapping alternative sources such as engaging in contract research. As a result, it has experienced a dramatic decrease in its total revenues; between 1990 and 1994, its income fell by one-eight in real terms.

In several countries we examined, an opposing policy was adopted by the Academy – that is, opting to make selective budgetary cuts. A key mechanism for deciding how such selective cuts should be made has been the introduction of evaluative procedures which have sometimes been supervised by foreign experts (see also Provaznik, Machleidt, & Filácek, this volume). The level of subsidy to particular institutes, and decisions on their fate (including whether they should simply be phased out) are typically linked to the verdicts of such performance reviews. However, some respondents expressed the opinion that, in practice, this policy did not always lead to the best institutes enjoying the protection they deserved since budgetary cuts in general were so severe as to be unable to guarantee the financial health and stability of worthy institutes. In the Slovak Republic, where reforms to the Academy system have been introduced incrementally, the evaluation of its institutes began in 1992. As a result of this initial review, the number of supported institutes actually increased. However, following the assessments of 1993 and 1994, they decreased again. While support for research in the Slovak Academy in 1990 was 14.6% of total R&D expenditure, by 1994 this proportion had fallen to as little as 9%. In that period, as a result of the combined effects of inflation, the devaluation of the local currency and the decrease in outlays on R&D in general, the financial circumstances of the Academy's institutes continued to deteriorate, although a slight growth was evident after 1994. In spite of the fact that staff reductions ran at about 45%, the outlay per researcher nevertheless decreased by about 28%. Even those institutes which had enjoyed the highest praise and validation as a consequence of the evaluation exercises suffered from budgetary cuts. For example, the Slovak institute of experimental pharmacology experienced revenue cuts of 60% in real terms over the four-year period of 1990 to 1994.

The Academy institutes in the Czech Republic have also been subject to assessment (half of them have currently undergone evaluation reviews) as have those in Hungary. In Poland, a continual mechanism for annual assessment exercises has been implemented and is used as a basis for decision-making about where resources should be allocated. At the same time, it has also been used as a tool for restructuring the whole field of research activity given that funding of science is on the decline in general. However, even those Polish institutes which have received the highest assessment grades have not received sufficient financial support to cover all their basic requirements. In other countries, the outcomes of review procedures implemented by the Academies have also been used as the basis for decisions on the allocation of funds but most typically the review procedure results in cuts to an institute's financial support. For example, the review of the Estonian institute of chemistry which was jointly conducted by the Academy and foreign experts resulted in a cut to that institute's funds.

Another type of policy pursued is to launch state programmes which set priorities on which kinds of research should continue to receive support based on criteria of social usefulness. In Latvia, such a national research programme was launched as late as 1995. In Belarus, a similar national programme was established in 1992 and was followed in 1995 by the Belarussian Academy of Sciences initiating its own research programme. In Romania, the financing of the Romanian Academy and its institutes between 1990 and 1994 was conducted within the framework of just such a national programme.

In addition to such priority-setting national programmes, another mechanism for determining whether a particular institute should receive support stems from priority-setting which has an overt political motivation. This is the case, for example, with the Lithuanian institute of language which is currently fully financed by government subsidy and which has a guarantee that it will continue to receive statutory funding for a further two years.

While many of the financial difficulties faced by institutes can be traced to issues of economic crisis during the period of transition, another major external factor on some institutes' financial hardship is explicitly political in nature. This is especially the case for the former republics of the USSR. The Academy institutes of the former Soviet Union had historically received subsidies from their own authorities and also from the Union's centralised Academy of Sciences. Following the collapse of the Soviet Union, the latter ceased to exist and an important funding source instantly dried up. In addition, these institutes had been used to receiving commissioned orders from the USSR ministries and from enterprises located in other republics. In particular, the industrial-military complex of the USSR had been an important purse-holder prior to the regime's demise. Those institutes which had tapped this source most strenuously were also those which suffered the steepest decline in income after the collapse of the political system. One of many such examples is provided by the Belarussian institute of physics. Prior to the collapse of the USSR all enterprises and institutes of the military complex were customers of this institute. Currently, however, orders from the defence sector are minimal and potential customers from other business sectors are too financially weak to be able to afford to commission work from the institute. Thus, the institute can only secure some short-term projects and the quality of commissioned research is on the decline. However, it remains the case that where a particular institute had enjoyed a monopoly on providing certain research services to the military complex prior to the regime's collapse, it subsequently became difficult for Russia to utterly discontinue the financial relationship in the period of transition.[2]

4. The Forms of Funding Institutes' Research and Their Changing Role in Institutes' Revenues

In all countries the major (but not exclusive) source of funds for institutes of the Academy was, and still is, the state budget. As already mentioned in previous sections, institutes are financed from the budget in the form of institutional subsidies.[3] Institutional funding is not a new form of income for many countries. However, the linking of the level of subsidy with the results of evaluation exercises is certainly an innovation in the Central and Eastern European countries' R&D systems. Another new form of state support is the provision of grants for individual scholars or research teams. These grants are distributed competitively and on the basis of peer review. In common with revenues raised by means of securing state contracts (i.e., commissioned research projects) from state ministries or private enterprises, grants are treated as an external and even an additional source of support for research activities. This does not imply that state subsidies are, or previously were, the dominant source of revenue for all establishments. Projects commissioned by industry were, in some cases, a major source of income. Since such customers still tend to be state-owned companies, their commissions remain important to some institutes. It must also be remembered that in many post-communist countries, these state-owned companies no longer possess their own in-house research laboratories and so they must contract out their research requirements to external research institutes.

While previously the majority of research projects which emanated from international co-operative endeavours were financed by COMECON programmes or were bounded within personnel exchanges within and between the

2 For example, the Belarussian institute of engineering had had just such a monopoly in its particular niche in the old regime. Thus, while the number of Russian orders dropped in the years following the ending of the USSR, they have not dried up completely.

3 With the exception of Latvia. In this country, grants predominate although 30% of funds are retained for centralised use to meet common needs.

Academies of Sciences of the Soviet bloc countries, this is not the case currently. Recent international co-operative research projects are funded or co-funded by the EC and NATO within a framework of various co-operative programmes and these new sources also offer foreign grants and foundational programmes of international exchange. Other forms of foreign aid include informal gifts of equipment, books or periodicals to institutes or to particular scientists working within them.

Other sources of institutes' revenue have nothing to do with research. Additional income is often generated by providing commercial or educational services, leasing buildings or equipment (the latter is particularly common), or interest earned on long-term capital deposits.

4.1 FINANCING BY INSTITUTIONAL SUBSIDY

The role of institutional funding as the basic source of finance can be seen in two ways. Firstly, it can be analysed as an absolute value which has altered over the period of transition. Secondly, it can be understood in relation to other forms of funding which, together, have undergone change in the restructuring of revenues.

In all our case studies, it was clear that the absolute real value of institutional subsidies has decreased. However, at the same time, its share as a proportion of institutes' total revenues often grew. This increase was especially significant in those institutes where the revenues from research projects commissioned by enterprises and ministries had diminished the most, i.e., those which had previously been oriented towards conducting applied research but which subsequently shifted towards basic research. For example, in 1990, 28% of the total revenue of the institute of informatics at the Ukrainian Academy of Sciences came from budgetary subsidy. By 1994 that proportion had risen to as much as 74%. At the same time the share of contract-work commissioned by industry (mainly from the armaments and aerospace industries) decreased from 72% to 9%. Similarly, at the Ukrainian institute of chemistry the total revenues in 1994 had decreased to one-eight of 1990's income in real terms, although the budgetary subsidy continued to account for 50% of the institute's revenues.

However, we also found an instance which suggests an opposite tendency, i.e., where the importance of the institutional subsidy decreased. In the Estonian institute of cybernetics 80% of its revenue came from budgetary subsidy (donations) in 1991 and 20% from contracts. By 1994 the subsidy only accounted for 44% of the institute's revenue. More than 54% arose from tapping into external funds.[4]

4 However, in the two other institutes of the Estonian Academy of Sciences which we examined, i.e., the institute of chemical physics and the institute of chemistry, the proportionate importance of the subsidy to the total revenues of those institutes increased.

4.2 FINANCING BY GRANTS

The importance of institutional donations as an expression of total revenues decreased slightly because scholars within the institutes could now apply for grants. This situation was observed in the Czech Republic, in Hungary and in Poland. In the latter, the success rate at gaining grants within two of the three Polish institutes we surveyed has continuously increased and is higher than the national average. These two particular institutes also received the highest evaluation grades – Category A. For that reason, they have also been successful in procuring a higher level of subsidy than institutes whose performance ranking is lower. Thus, these two institutes belong to a privileged group. However, this did not prevent the staff we interviewed there from expressing their discontent with their institutes' level of subsidy, since the value of that subsidy still cannot match the rate of inflation.

A system of grants which are made available for individual researchers and teams has been introduced in all the post-communist countries. Grants are allocated competitively or from within the framework of the Academy's procedures and/or by agencies and foundations which have been specially appointed for the purpose. Since grant award schemes constitute the most important innovation in the financing of scientific activity, some modes of grant distribution in post-communist R&D systems will now be described in a more detailed way.

In Romania, upon the transition from predominantly financing institutes by means of institutional subsidy (which are remitted to research programmes), the Ministry of Research and Technology and The Centre for Research on Higher Education located within the Ministry of Education are the bodies responsible for the competitive allocation of grants. In Latvia, grants are allocated by the Science Council and this body is also the agent which directly finances institutes (in this country, only 30% of funds are allocated on a centralised basis). In Hungary, the grants allocated by OTKA and OMFB play an increasingly significant role. Recently, the latter agency has stipulated that grants awarded for applied research projects must satisfy OMFB's system of revenue returns; they require that companies must be able to implement or manufacture the outcomes of the grant-aided research conducted by the institutes.[5] Agencies allocating grants have also been established in other countries and significant importance is attached to them in the Czech and Slovak Republics and in Poland. In the Slovak Republic, two separate agencies have been established which disburse grants either for science or for technology. Latvia and the other Baltic countries also operate grant-award schemes.

The role of grants as opposed to institutional subsidies within the surveyed institutes depends on several factors. First, the grant system can be channelled

5 Unfortunately, these new provisions have had a negative impact on institutes' interest in tapping into this kind of funding support.

through a pool of monies which is discrete from the funding pool for budgetary subsidies. For example, an Academy of Sciences may operate its own distinct budget and allocate grant funds according to its own criteria. Alternatively, as is the case in Poland, grant awards may be sourced from the same funding pool which directly subsidises institutes' revenues. Under the latter distributive system, any increases in the number or monetary value of awarded grants have a direct and immediate effect in reducing the monies available for subsidy. For example, the funds for financing the Academy's institutes fell in the Slovak Republic by 20%-30% following the launch of the grant award scheme in that country. In essence, this example illustrates how the grant system may simply operate as a mechanism for re-allocating the same (or diminishing) amounts of funds rather than representing a genuine alternative source of revenue for hard-pressed institutes. Because of the general decrease in the total amount of monies available for financing research activity, institutes may lose part of their revenue under the grant allocation scheme unless their employees succeed in winning an adequately high number of grants to compensate for the loss.

Second, the role of grants versus institutional subsidy depends on the importance that a country's science policy attaches to these different forms of support, as revealed by how that policy translates into financial proportions. The national science policy influences decisions on how many grants are awarded, what total monies are available to support grant applications, and what the average value of each grant is. This last issue assumes particular importance when there are more widespread budgetary difficulties. For example, both Hungary and the Slovak Republic display a great interest in the grant scheme and there has been an increase in the number of grants awarded as a consequence. Nevertheless, despite the sheer number of grants awarded to institutes in those countries, the importance of grants as a proportion of the institutes' total income has only very slowly increased since the value of the average grant is relatively small. In Poland, grants constituted about 18% of the budget in 1991 but, in line with budgetary constraints on scientific activity, that share fell to about 14% in 1994 because the maintenance of research potential and statutory financing were considered to be the most important policy goals.

Third, it is important to understand how the grant allocation policy works in a context where funds for grants are held at a constant although the number of applications keeps increasing. In many countries we hear complaints that individual grants are rather small and therefore have no real significance for augmenting the institutes' revenues. Given that institutional subsidies in most countries barely cover the cost of infrastructural maintenance of the establishment and of meeting employees' salaries, the role of grants is increasing. They become the basic source for financing the direct costs of research.

Fourth, a crucial issue is the amount and kind of costs which the grant is intended to cover. Individual grant schemes adopt differing criteria on how overheads, salaries and other costs may be included in, or excluded from, the grant provision. What is the acceptable margin of overhead costs which the institutes may claim from the grant and what proportion should remain at the disposal of the grant-holder? There can be a conflict here between the interests of the institute and the individual grant-holder. An institute may strive to recover all its overhead costs from the grant while the grant beneficiary tries to retain maximum leverage so as to channel the grant funds directly into meeting the immediate costs of the research.[6]

In most countries, grant money can be spent on providing additional income to the research staff who are carrying out the project. However, sometimes there are limits set on the extent of such honoraria. For example, since 1995 in Romania, money from grants cannot be spent on salaries. In the case of funds generated from contracts, up to 70% may be spend on equipment and 20%-30% on meeting personnel costs – although honoraria for researchers may not exceed 50% of their basic salary. In other countries, slightly higher levels are allowed. In Bulgaria, half of the funds raised from contracts may be spent on maintenance of the infrastructure (i.e., buildings), a quarter on salaries, and a quarter on the infrastructural costs of maintaining departments and meeting travelling costs.

Fifth, the role of grants versus institutional subsidy depends on the kind of activities which scholars intend to undertake in order to meet the criteria for winning the grant and on their success in meeting their chosen projects' targets. In some countries, the number of grants awarded to the institutes we surveyed was on the increase, as was their success rate in meeting the achievement criteria set by those grants.

Sixth, the existing systems of grant disbursal can actually work to restrict the kinds of research activities carried out by the institute. In some countries – Hungary and Poland, for example – grant applications may only be submitted by individual scholars and not by their institutions. In the Slovak Republic, the grant system operated by the Science Agency has made it impossible at times for institutes to conduct joint projects with scholars from higher education establishments, since higher education grants are financed separately from a different grant award scheme.

The role of institutional financing versus grant-based financing is one of the most broadly discussed problems across all our case studies. The importance of institutional financing is emphasised, since it allows for infrastructural development of the institutes. In some countries (e.g., Latvia) our respondents noted that the method of organising the distribution of finances in the form of grants results in a deterioration of the infrastructure. Supporters of the budgetary sub-

6 In the Latvian system, 40% of each national grant devolves to the institute to meet overheads while 10% is creamed off international grants and contracts for institutional purposes. In the Slovak Republic the overhead on grants is 20%-30%; in Bulgaria, 10%; in Hungary the overhead on OTKA-grants is 17%; in Poland, 20%. In other national schemes, the proportion allocated to meeting overheads is usually negotiated on a contract-by-contract basis.

sidy allocation mechanism were especially prevalent in those institutes which had, by tradition, specialised in the provision of basic research and in those institutes which had had to re-orient themselves towards conducting basic research to compensate for the fact that the changed external environment meant that commissioned contracts for applied research had grown scarce.

It must also be remembered that grant systems which allocate monies to individual scholars rather than whole institutes imply an additional hazard to the general autonomy of those institutes. The grant system is based on the freedom of research choices at the level of grant-holders, and those project choices need not correspond with the research goals of the institute as a whole. Where grant schemes replace or significantly displace direct budgetary subsidies to institutes, those institutes become restricted in their ability to co-ordinate and conduct their own independent research policy and to formulate strategies for future development.

Emphasising the continued importance of institutional financing does not imply a negative attitude towards grant schemes. It was stressed by our respondents that grants awarded to individual scholars enliven the research community; reward personal initiative; offer individuals significant freedom in framing their research; and allow the best scholars to thrive. In general, one may state that a combination of both forms of financing enjoys widespread support – providing that funds for research are sufficient. However, in practice, both institutional subsidies and the average values of grants are too meagre and thus neither can be managed and held in balance in a reasonable way.

Aside from Latvia (where grants constitute the basic form of financing of the Academy's institutes), revenue raised from domestic grants account for only a small share of total income, ranging from several per cent to more than 10%. Since new opportunities have arisen to pursue foreign grants and openings now exist to participate in international co-operative scientific programmes, these income-generating routes are assuming an increasingly important role for institutes located in Central and Eastern Europe. Already, foreign grants frequently contribute more to the total revenue of institutes that do domestic grants.

5. Changes and Adjustments Within the Institutes

Previous sections have described how the transformation period has introduced new conditions for how research institutes may function. What kind of changes can be observed? How Academy institutes now operate in their altered financial circumstances is a consequence of how two kinds of factor are linked.

The first set of factors has already been discussed, i.e., externally-imposed changes concerning the scope and availability of resources, the methods and structures of financing, and the criteria for funds allocation. The second factor refers to how institutes themselves are adjusting to their changed circumstances in terms of their own internal financial management practices. Within any given institute we may find a number of patterns of adaptation. In order to illustrate how these two kinds of factor shape the tactics that individual institutes deploy, we will briefly describe how four institutes from the Russian Academy now function.

Although all four institutes of the Russian Academy of Sciences have suffered from a deterioration in their financial well-being and all report that their share of budgetary financing is insufficient (at present, this constitutes about half of the institutes' revenues) the institute of theoretical physics is relatively in the best situation. For instance, it can still afford to employ young scholars and our respondents note that it maintains a high scientific standard. The Institute carries out projects within the framework of priorities set by the Ministry of Science and Technology Policy. Researchers are awarded grants from a Basic Research Fund. Besides that, all its scientists participate in various international research programmes which help to generate additional revenue. In contrast to the other three institutes, the employees of this establishment receive their basic salaries on time. However, the condition of the infrastructure is poor; the institute does not have its own buildings, and the equipment used in some of the research specialities pursued here is obsolete relative to that at the disposal of the scientists' international peers.

The financial health of the other three institutes – the institute of Economics, the institute of radio engineering and electronics and the institute of biology – is far worse. The institute of economics, which used to be "meaningful for science" according to our respondents, currently operates in a highly competitive environment and must pit the services its offers against other newly-established entities. The institute of radio engineering and electronics used to work extensively on behalf of the military- industrial complex but it has now lost this source of financing. Its director estimates that its level of financing has decreased to approximately one-thirtieth of the level it enjoyed in 1988. Staff levels are down in all these institutes – although in two cases, staff reductions have only affected non-research personnel. However, the institute of biology has also had a proportion of its leading scholars made redundant.

These three institutes are seeking opportunities to secure funding from other sources – just as the institute of theoretical physics has done – but they face a different set of obstacles to their favoured sister institute. The first difficulty pertains to co-operative projects with business since, in many cases, those enterprises who might be potential customers or partners are insolvent. The second difficulty concerns how they might profit from renting out their premises (all three have got their own buildings). Renting out their premises is not entirely legal since the buildings are controlled by the Presidium of the Academy and the institutes "get nothing out of it". Therefore, one of them has simply side-stepped the regulations and is letting out space under the guise of pretending to be in a co-operative relationship with their tenant.

In all three of these institutes reduced institutional funds are remitted in monthly instalments, are often delayed, and the financing is "unstable and not reliable". As a result, it quite often happens that there are no funds to meet salaries – this has occurred at both the institute of economics and the institute of radio engineering and electronics. In addition, there are 'dead' posts; compulsory leaves of absence have been implemented; payments due to employees are delayed; and, periodically, shorter working weeks (e.g., 3-day weeks) have been introduced. In the institute of biology salaries are paid irregularly.

Institutions obtain funds and scientists supplement their income by participating in international programmes. The institute of biology, in addition, operates joint projects which are co-financed by the Russian government. In all three institutes complaints are voiced about inadequate equipment and poor technical standards – notwithstanding that their computer networks have been modernised. In the institute of radio engineering and electronics, no new equipment has been bought for the last eight years. Therefore they do not conduct any practical experiments and their current research activities are largely confined to theoretical projects. Purchasing equipment and materials from abroad is subject to high excise duties. Thus, it happens that scholars smuggle items across the border, as was reported to us at the institute of biology.

Some revenues are derived by the institutes (and by individual scholars within them so as to supplement their salaries) from running independent commercial enterprises. These have been set up, for example, at the institute of radio engineering and electronics and the institute of biology. However, the income generated from this kind of source is rather meagre.

The financial situation within each of these institutes differs. The changes to the system of research financing and fund management has led the institutes towards a situation where individual units and research teams pursue commissioned projects and grants on their own initiative and to meet their own interests. These individualised initiatives operate separately from the institutional subsidy which is distributed and allocated by the institutes' administrations.

The difficult financial conditions of the institutes under consideration here have resulted from many factors. However, the particularly straitened circumstances of the institute of radio engineering and electronics has been aggravated by the fact that the institute continues to maintain expensive and over-developed employee welfare facilities which are located in the institute's subsidiary outside Moscow. These facilities include three kindergartens, a pioneer camp, a canteen and some housing stock. The institute does not want to get rid of these facilities and hand their welfare functions over to the municipality, "because everything would be finally ruined".

These examples, drawn from just one country, illustrate the variety of response within institutes regarding the structure of expenses. As a result of that variety, we also find differences in labour force management, in how each institute manages its material, organisational and social affairs, and in how it regulates the conditions for carrying out research work. When our comparison stretches to cover institutes from various countries the differences between the varieties of internal transformation are even more striking.

5.1 IMPACT OF FUNDS REDUCTION ON HUMAN RESOURCES AND SALARIES

When studying the management of scarce financial resources we note that most institutes prioritise some expenses as immutable and others which they regard more flexibly. These latter costs are the ones which they seek to reduce, or dispense with, for a short period of time. The former group of costs includes maintaining and protecting human and material resources. They refer to the elemental expenses of meeting basic salaries, maintaining buildings, libraries and utilities, and providing basic communal services. The latter group which is set at a lower priority includes investments (e.g., the construction and repair of buildings or the purchase and repair of instruments) and the sundry expenses associated with the direct costs of research (e.g., materials, reagents, travel expenses, copying materials, conferences, etc.).

How institutes prioritise their various categories of expense can be seen implicitly in the way they have presented their financial difficulties in their national reports. All have tended to highlight their efforts to meet salary and infrastructural maintenance costs from their reduced revenues. The implied ranking of priorities is probably influenced, amongst other factors, by the structure of the various forms of funding they receive (i.e., the institutional subsidy is typically the dominant funding mechanism) and the administrative or legal regulations concerning how funds from this source may be deployed.

The various internal institutional reorganisations we observed were shaped by the institutes' legal and administrative framework of financial management. This framework – either directly or indirectly – worked to enhance certain kinds of institutional response and to set limits on the institutes' flexibility of response to other issues such as their institutional autonomy. In some countries, pre-set regulations concerning the maximum level of overhead to be creamed from grants and contracts; payments for participation in grants or contracts; or the way the basic salaries of researchers are established (e.g., whether researchers' pay is linked to that of civil servants), can all become obstacles to how flexibly an institute's managers may respond to the institute's new circumstances.

What kinds of reaction could be observed amongst the institutes we studied as they grappled with the serious constraints on their spending power? Prior to presenting some general patterns of adjustment, it is useful to compare responses to the single most protected set of costs – that of managing human resources – so as to illustrate the various consequences the financial cuts have had on institutes' adaptation patterns.

Faced with cuts and/or a decline in the real value of subsidies, institutes have had to choose between two basic options; they may either manipulate salaries or they may adjust their employment levels downwards. For most, the former tactic was not really available to them; salaries were already set at a minimum acceptable level, below which all their best employees would leave the institute, where they really had an exit option. Thus, for most, it was easier to dismiss some of their employees.

In most cases, we found a decrease in employment levels at the Academies' institutes between 1990 and 1994/5 which most of our respondents directly linked to funding constraints (see also Provaznik, Machleidt, & Filácek, this volume). We can discern two basic patterns of adjustment behaviour, one being passive/reactive and the other being active/anticipative in nature.

In some institutes redundancies were made only after the institutional subsidy was reduced. These staff reduction measures were the institutes' reaction to financial cuts, since they were not able to cover basic salary costs. However, institutes in those countries where verification or assessment exercises of research bodies were conducted tended to react differently. Quite often, they carried out their own internal review and evaluation of their employees and selectively let go their less productive staff before any external review took place. As a result they did not have to suffer quite as large a reduction in their funds as was the case for those institutes who were still employing poor performers.

Staff reductions in most institutes affected a higher proportion of releases amongst auxiliary and administrative staff than amongst researchers. Again, however, we find two kinds of strategic reaction. The first our interviewees called "the protection of the intellectual research potential" of the Academy (regardless of the real quality of its employed researchers) and this strategy was pursued by institutes which adopted a predominantly defensive stance. When we compare falling employment levels with declining subsidies, we can see that the defence of 'scientific potential' was only partly effective. This was because such defensive actions produced side-effects. Following the dismissal of administrative or technical staff whose salaries were lower than those of scientists, the basic salaries offered by those institutes were maintained at a low level, wages remained undifferentiated, and shorter working days were introduced. However, in some of the institutes employing high-calibre scientists, the release of some categories of employee were the result of a pro-active policy implemented by the institutes' managers which was aimed at ensuring financial savings. They rid themselves of those administrative and auxiliary posts which, given their newly-autonomous status, were no longer relevant.

It must be mentioned here that some countries link the level of subsidy either with the total number of employees or with some particular categories of staff. In these cases, it was more difficult for institutes' managers to establish a reasonable human resources policy. The continued linkage between subsidy level and staff numbers did not work to promote initiatives to reduce staff numbers. In addition, it was sometimes the case that despite reducing staff numbers, the external regulations governing the level of subsidy permitted per employee remained unchanged. Frequently, increases in the subsidy could not match cost-of-living increases or costs associated with the provision of communal services.

The other problem which the institutes' managers had to face was fixing salary levels in contexts where those levels did not solely depend on the institutes' own resources. For managing the salary issue, it is important to establish whether, and to what extent, the institute's authorities may independently determine the level of basic salary. In some countries, institute staff are regarded as civil servants while, in others, the exact salaries commanded by particular categories of post are pre-determined centrally. Sometimes the institute's board might augment the salaries of particular staff members by means of bonus payments. In practice, however, these extra top-ups are typically modest. Thus, although an acceptable practice, selectively raising the financial package being offered to some staff above the level of the basic salary is only a limited measure. Typically, it is only done by institutes which are actively attempting to stimulate individual scientific productivity.

As basic salaries are low, staff often attempt to supplement their income by applying for grants and contracts or for participation in foreign programmes. Many engage in educational activities or consultancy work or they write commissioned reports in their area of expertise. Institute managers who fall into the active/anticipative category typically encourage their staff to augment their personal income in ways such as these by, for example, rewarding those who raise additional monies for themselves or their departments. As this encouragement does introduce diversity of income amongst similarly-graded employees, other managements go to the other extreme and aim to encourage complete equality of income amongst similar staff categories.

We have seen that some of the most common changes in employment practices as a result of the transformation process within institutes include a gradual lowering of salary levels; a general reduction in the additional income available from contracts; delays of payments due to employees, sometimes including their basic salaries; the imposition of part-time work; and, compulsory leaves of absence. In these hard circumstances, it is not surprising that so many institute staff seek to earn additional income within the research or educational fields. However, many staff also double-job in areas unrelated to their core expertise (e.g., in manufacturing) or they set up their own entrepreneurial companies while remaining employees of the institute.

In most institutes, it was common for staff to undertake additional employment and most managements do not discourage double-jobbing. As a result, the performance effectiveness of researchers has tended to suffer as their commitment of time and energy to the research task is stretched thin.

The financial problems described above have produced a brain drain of scholars who are either leaving to go abroad or to enter other sectors of the economy. In addition, financial constraints have slowed down the in-flow of young scientists to the institutes (see also Nesvetailov, this volume). There are few easy solutions to these problems. Even one of the most promising new opportunities – that of participating in international research programmes – has a downside. Many of the most promising scholars who avail of foreign junior positions or scholarships do not want to return. However, many have been tempted back in those countries which offer special financial incentives or financial aid to returnees for a period of re-adaptation.

To sum up, we can distinguish some common reaction patterns amongst institutes facing financial restrictions. We have classified them according to whether managements adopt a passive or a proactive stance towards handling change, including how they acquire new sources of income.

First, we found that some instutites orient themselves towards defending the status quo, i.e., they pursue an egalitarian approach. They aim to protect the entity as a whole without carrying out significant internal changes. In such cases, staff reductions are minimal; funds generated by the most active groups are re-distributed within the institute; grants are used to cover both overheads and other general running costs such as postage, xerox paper, travelling costs, and outlays connected with visits by foreign scholars. Typically, these institutes are simply waiting "for better times".

Second, passive forms of adaptation tend to produce internal institutional outcomes such as reducing employment levels as, and when, funds are cut back; scaling-back salaries; and selling buildings to raise revenue. In these institutes, the scope of research activity is also reduced because of a lack of funds.

Third, institutes pursuing a pro-active strategy seek to make savings before any such cuts are forced upon them by external circumstances. They actively look for ways to reduce their financial burden, including making staff reductions and leasing out unused premises. Sometimes, these institutes also introduce financial decentralisation procedures.

Fourth, these anticipative managements orient themselves to selectively protect the strongest parts of their institutes. Typically, they carry out their own regular internal assessments of their departments and of individual staff posts. On this basis, they make internal decisions on selective reductions. The best and most active units are rewarded (e.g., by means of salary differentiation) and individuals and units are stimulated to tap alternative sources of income or to initiate their own commercial start-ups.

Fifth, anticipative action tends to produce internal restructuring to increase future chances to obtain additional resources by both individual staff members and by the anticipative institute as a whole. Typical actions include applying for grants and seeking out new contracts. Endeavours to secure foreign funds are strongly stimulated and rewarded. Such institutes also attempt to widen their general profile to include new forms of income-generating work activities. Regular internal reviews and assessments of departments and staff are accompanied by productivity reward schemes (e.g. promotion or bonus schemes) and thus the internal distribution of the institute subsidy is linked to internal achievement criteria. These institutes also tend to implement financial policies which are actively geared towards protecting the continued in-flow and job security of younger staff members.

The institutes which are most successful in coping with financial crisis are those which apply the most radical pattern of active management. However, these institutes are not always the same ones which have been a particular nation's traditional centres of excellence. In addition, only very few institutes have actually forged such a radical management path. The majority of institutes cannot, in practice, be characterised as belonging to the pro-active category described here. Since diminishing resources to support research activity is a normal situation nowadays (and not only for countries undergoing post-communist transformation) the future of many research institutes is still at issue.

COPING WITH SCARCITY: STRATEGIES OF PERSONNEL REDUCTION

STANISLAW PROVAZNIK
PETR MACHLEIDT
ADOLF FILÁCEK[1]
Centre for Science, Technology, Society Studies
Institute of Philosophy
Academy of Sciences
110 00 Prague

1. Introduction

An international comparative analysis of studies relating to staffing issues in research centres within the post-communist countries is, in general terms, difficult to attempt. Such an analysis requires a knowledge of the wider social, political and economic contexts within which research is being conducted during the period of transformation together with an understanding of the historical specificities of these contexts across the different transitional societies of Central and Eastern Europe. In this respect, staffing changes in the research systems of different countries are assumed to depend essentially on two factors:

- understanding how radical the political and economic reforms in specific countries are, and
- understanding how deep the economic depression has gone within specific countries as a consequence of implementing reforms.

Together, these circumstances largely determine the strategies for action which have been chosen by specific scientific communities during the transformation process.

These circumstances must influence our exposition and analysis of the issue of staff reductions in the post-communist research systems we shall be looking at. Upon examining the case studies, we consider that similarities and variations amongst the different countries in Central and Eastern Europe testify

1 With the assistance of Eva Krizova-Frydova and Jiří Loudin.

R. Mayntz et al. (eds.), East European Academies in Transition, 33–47.

to the fact that each country is following its own unique path towards reforming its system of science and research. It would be illusory to try to find a generally applicable recipe which fits the settings of all the countries we have analysed. Thus, our analysis should be understood as a suggestive pathway towards gaining a more complete understanding of the difficulties being encountered by the reforming countries which, in turn, provides the contexts within which the general problem of the considerable reduction of research staff is located.

In the following sections of this chapter, we will first describe the structure and dynamics of staff reduction in post-communist R&D systems, based on the data presented in the case studies. Then we will turn to the commentary provided by the experts who were interviewed in the course of the research in order to tease out the ambivalence of the responses to the issue of staff reduction. Next, we show the correlations and links between staff reduction and other recent developments in the transformation processes of the post-communist R&D systems. Finally, in our concluding remarks we highlight two different approaches to the strategy of staff reduction which can be discerned in the R&D transformation processes within the respective countries. The first of these approaches is a defensive survival strategy. The second is a rigorous reduction strategy which is guided by scientific efficiency criteria. We conclude that it is only the latter strategy – as implemented in the Czech Republic – which provides a successful example of the transformation of R&D systems in Central and Eastern Europe.

2. The Dynamics and Structure of Staff Reduction

Staff reduction in the R&D system is underway everywhere in Central and Eastern Europe and it constitutes part of the transformation of these societies as a whole. Along with political and economic transformation processes, staff reductions within the Academies of Sciences began in the mid- to late-1980s (e.g. in Poland and Hungary) although, in most countries, the highest reduction of research staff took place between 1990 and 1994.

2.1 THE EXTENT OF STAFF REDUCTION

Based on available data, the rate of reduction from 1990 to 1994 in the Academies of Sciences as a whole across the different countries (that is, not only in the institutes selected by the study for particular consideration) oscillated between 10% and 53%. The average was approximately 33%. In the institutes we observed in our study, the extent of the reductions was, in general, lower (at 23%), although the staffing levels in some of them decreased by as much as

50%.[2] In other institutes, staffing levels did not fall by more than 11% and in some cases staff numbers were actually increased rather than cut. Nevertheless, our respondents believe that in these particular cases, staff reductions have simply been postponed and that the numbers will fall in the coming years.

Staff reductions in the institutes we examined were realised in two ways, either as a deliberately organised reduction policy or as a more individualist and spontaneous series of resignations by particular employees seeking better paid positions. The latter case describes a permanent and small-scale process of normal turn-over and is associated with younger and more capable staff. It is difficult to conjecture the ratio between the two forms of staff reduction. According to the data from our case studies, voluntary resignations account for approximately 10%-30% of the total number of staff leaving the institutes. For instance, in the Polish institute of biology (for staff engaged in molecular biology and genetics) 10 out of the 40 staff members left spontaneously in 1994.

In all the countries we analysed neither the completely planned and regulated route towards staff reduction nor the spontaneous resignation route occurred in isolation from each other. In most cases, we found a mixed compromise between these two extremes. However, the science community itself has not always had an opportunity to influence the process of staff reduction. A high level of scientists' participation in the staff reduction process was characteristic of the institutes in the Czech Republic however. Here, politicians and administrators gave the scientific community the power to guide and supervise the transformation in a relatively autonomous manner. Thus, staff reductions have been less painful in these institutes than in other post-communist R&D systems, although the strategic pursuit of staff reductions has not been less principled or determined.

As far as the different research branches in our sample of Academy institutes are concerned, the greatest reductions have taken place in the institutes of physics while the institutes of chemistry have escaped relatively lightly. Minimal reductions took place in institutes dealing with boundary fields (e.g., biophysics and biochemistry) which may be explained by the greater involvement of these research fields in international co-operative endeavours, the better provision of international grants for such work, and the superior contacts such research enjoys with industry and business. In addition, the opportunities to work abroad are better in these research fields with the significant consequence that outstanding scientists may choose to emigrate.

The data on staff reduction may sometimes obscure the issue of quite considerable staff mobility. For example, in the Ukrainian institute of chemistry, 80 employees were made redundant and 100 new staff hired in a single year. In the Polish institute of biology (molecular biology and genetics) employee numbers

2 Indeed, the Belarussian Academy of Sciences reduced its staff by 60%.

remained roughly constant at approximately 200 staff between 1990 and 1994. However in 1992 – the year of greatest turn-over – 40 staff left while 41 new employees were hired. Virtually no reduction took place in any of the three observed institutes of the Romanian Academy of Sciences (Physical Chemistry, Theoretical Informatics and Biochemistry). Changes here were realised by means of staff mobility while the total number of employees remained stable.

The structure of staff reductions can be considered from several angles. Firstly, in the majority of institutes the reduction process primarily affected auxiliary rather than research staff. In some cases, this tendency was highly marked. For example, in one institute research staff numbers were reduced in 1990-1992 to 82% of the previous level whereas auxiliary staff numbers were slashed to 22%. In another institute, the number of research staff actually increased in the same period to 113% while auxiliary staff numbers shrank to 37% of the previous levels. The number of auxiliary staff has fallen significantly in the institutes of the Slovak Academy of Sciences. The share of research staff in the total number of employees has also considerably increased in the Romanian institute of physical chemistry (from 12 staff members in 1989 to 47 in 1994). In Estonia, however, the balance has remained essentially the same. The data lead us to conclude that the motive behind the reduction of auxiliary staff in many institutes was to protect the highest possible number of research posts. Sometimes, indeed, auxiliary staff redundancies have had the effect of worsening the technical provision of the research itself (see also Jablecka, this volume).[3]

Secondly, administrative staff numbers have not been cut as drastically as auxiliary staff numbers. In some cases, administrative positions have even enjoyed greater security than research posts; our study reveals two cases where administrative staff numbers actually increased between 1990 and 1994. A likely cause of this astonishing phenomenon is the process of decentralisation within the Academies whereby administrative duties are being delegated from the central managements of the Academies to the level of local institutes.

A third angle on the structure of staff reductions refers to the fact that, amongst researchers, research staff with the highest scientific degrees are less affected by reductions than are more junior researchers. Doctors of Science (Professors) possess the greatest power to defend their existing posts. This tendency could be observed in virtually all the institutes were this index was inspected – that is, in the institutes of Bulgaria, Russia, the Ukraine and the Slo-

3 There are exceptional cases, though, in which staff reductions have brought about both absolute and relative decreases in the number of research staff. For instance, in the Ukrainian institute of informatics (cybernetics), the proportion of research staff to the total number of employees decreased from 58% in 1990 to 46% in 1994.

vak Republic – where the number of research staff in this senior category has increased absolutely by an average of one-third in recent years.[4]

In addition, staff reductions have affected the social sciences and humanities more than the natural sciences. This may be explained by differences in the extent of the ideological burden carried by the social sciences/humanities as compared with the natural sciences. For example, the institute of philosophy in the Czech Republic experienced a change to 75% of its staff within two years of the overthrow of the previous regime. However, an exception to this tendency can be found in a Latvian language institute where employee numbers actually increased, the reason apparently being the politically-motivated need to rediscover and develop the national language as an element of Latvian nation-building. Considerable staff reductions in the natural science institutes have only been implemented in those organisations which have fundamentally changed their thematic orientation and such institutes are in the minority.

At the outset of the transformation process, worries were frequently voiced that a mass of young and talented researchers would emigrate and thus damage long-term national research potential. These anxieties have proved to be unfounded, however. For example, within the entire Russian Academy of Sciences only 0.8% of researchers (i.e., 508 staff) left the country in 1991-1992. On the other hand, some of the scientists who had emigrated in the 1970s and 1980s actually returned to their institutes in the 1990s (this was especially true of Poland).[5]

More typical than wholesale emigration has been the tendency for research staff to undertake long-term international visits (see also Mirskaya, this volume). In some institutes, especially in the Czech Republic and Hungary, as many as 25% of research staff are currently on international trips of greater than 6-month duration. The length of international stays is, however, gradually shortening. In one of the Hungarian institutes the average length of an international stay amounted to two or three years in 1990; by the mid-1990s this time period was reduced to an average of no more than one year. The supply of international invitations to visit the Western countries is declining and the demand is influenced by increasing mutual competition amongst young scientists from the Central and Eastern European countries. However, not all scientists who return from international stays continue their research work at home. In many cases they leave the R&D system completely.

4 Other countries did not survey this index. It may be that they did not consider it relevant for the evaluation of the quality of their research staff or that they accentuate different methods to assess the credentials of their research staff.

5 The returning scientists indicated the following reasons for their return: a greater independence of research choices available to them in their home countries than in the West, less competition, greater stability and security, and better opportunities for self-fulfilment.

In the course of staff reduction processes, an intolerable number of the promising young scientists have left the R&D system (see also Nesvetailov, this volume). Case studies unambiguously confirm the well-known fact of the declining interest of young people in research jobs, especially in the natural sciences. In the observed institutes the number of research staff under 35 years of age has approximately halved. There are virtually no staff under 25 working in the institutes. However, there are exceptions: for example, young people continue to be attracted to the Russian institute of theoretical physics. It seems they are drawn by the institute's excellent reputation and its numerous international contacts which offer them the means to travel abroad. The exception among Academies is the Romanian Academy of Sciences where 20% of researchers are under 30 years of age. In the institute of biochemistry, which is not typical of the Romanian Academy as a whole, the number of young staff has increased by 40%.

2.2 EFFICIENCY AS A CRITERION FOR STAFF REDUCTION

In respect of research efficiency, who has been leaving the academic institutions? In all the observed institutes where staff reductions have taken place, it seems that the aim has been to reduce the number of less efficient researchers. In questionnaires which were distributed to selected institutes, respondents were asked to classify job terminations into separate groups according to research credits and efficiency. In many institutes the proportion of senior or productive scientists increased slightly, whereas the proportion of less productive staff members significantly decreased. In the interviewees' opinions, in most cases staff reductions have left only a few scientists in the R&D system whose productivity remains questionable. For example, in the observed institutes in Russia, it was primarily the less efficient members of staff who left during the staff reduction process; respondents estimate that the intellectual potential of the institutes is higher after transformation than it was previously. However, it was also pointed out that in one of the observed institutes very capable and efficient researchers of young and middle age were leaving in addition to less qualified members of staff. Their number reached 50%-70% of all job terminations. It is only in the case of some of the institutes in the Czech Republic and Poland that the case study data shows that, as an effect of staff reduction, the research capacity of research centres has not decreased but, on the contrary, has increased or "somewhat increased" as some interviewees point out. The evidence for this argument is predicated on the growth of publications in leading international journals with a high impact factor (see also Schimank, this volume). For example, at the Polish institute of biology the number of publications has grown as follows:

	Publications
1991	92
1992	141
1993	135
1994	128

The number of citations according to the SCI was as follows:

	Citations
1991	849
1992	1135
1993	1330

In the Czech Academy's institute of physics the number of publications in the international scientific journals was as follows:

	Publications
1989	206
1990	265
1991	254
1992	257
1993	284
1994	169

The number of citations (self-citation excluded) in the databases of SCI was as follows:

	Citations
1989	1230
1990	1079
1991	1329
1992	1838
1993	1906

The interviewees from most of the institutes which have experienced staff reductions believe that their organisations have succeeded in saving the core of the research potential in their respective institute. More than half consider that staff reductions might, to a certain extent, continue into the future without essentially violating the research efficiency of their institutes.

2.3 VOLUNTARY STAFF REDUCTION BY JOB MIGRATION

The reasons why qualified and efficient research staff leave for other professions are the low salaries of the research professions together with their decreasing social prestige. Employees of institutes who had been made redundant usually found new jobs outside the research sector. They transferred mainly to public administration, banking, and private (especially international) businesses. They were attracted by higher job security and better financial rewards in these sectors. Ex-staff members with proven expertise and foreign language proficiency have not had significant problems finding new jobs in most countries. Job and manpower transfer within the Academy has occurred more rarely; transfers to universities were more frequent, although they only concerned those researchers who worked in research fields which were formerly under-represented in the university framework. Staff seldom left to engage in applied research: opportunities for such transfers were limited to certain technological fields. Those researchers who left the research sphere, especially those entering business and banking, did usually not return to research.

3. Ambivalent Consequences of Staff Reduction for the Transformation of Science: Two Different Points of View

During the transformation of the R&D system the Academies of Sciences in all the countries in Central and Eastern Europe have undertaken comparatively large staff reductions, which have no parallel in their previous histories. For the last few decades, these countries – most of which were, and are, economically under-developed – have built up numerically strong and large scientific and research potential and ascribed to it a special value in their cultural and national sense of self-worth and state pride. Today the post-communist societies are confronted with the necessity to seriously examine and revise this potential over a very short period and in accordance with the criteria of efficiency, performance and social utility. Thus, the cultural elites must cope with enormous difficulties in evaluating their native R&D systems.

If we leave aside the significant existential impact of the transformation on the fates of particular people and focus only on the effect of that process on academic science, we are confronted from the outset with two completely opposing views held by scientists as they evaluate the transformation's course and effects. Only in a small number of observed institutes (and only in some countries) do scientists themselves consider staff reduction in Academies of Sciences to be a positive phenomenon. Some of these scientists view staff reduction as the essence of successful transformation, since such pruning eliminates excessive and inefficient research capacity, terminates the jobs of less efficient research staff, and improves work conditions for those who are more productive in their research.

Nonetheless, in a high number of research institutes in many countries we also encounter much more pessimistic evaluations. According to the scientists who view the situation more sceptically, ongoing staff reduction considerably weakens the extant research potential of the nation which, in turn, leads to a general decline of science. The argument runs that successful research teams fall apart, that talented young scientists leave research, and that many of those remaining gradually lose interest in working in a general atmosphere of decline and decay. The transformation processes of science systems in the countries of Central and Eastern Europe are everywhere characterised by their complexity. Therefore, it is exceptionally difficult to address whether current research staff reductions affect the status and development of science positively rather than negatively, or to what extent such reductions contribute to the successful transformation of science rather than to its deterioration. The answer depends on a number of relationships between the system of science and the rapidly changing social and economic context.

4. Staff Reduction as an Essential Feature of the Transformation of Post-Communist Research Systems

The necessity for staff reduction was indisputable from the outset of the period of transformation. None of the Central and Eastern European governments that were established in the period of the setting-up of market economies in place of the traditional command economies, had sufficient financial resources to continue to support institutions as large as the Academy of Sciences without amendment.

In this respect staff reduction represents one common reaction to the necessity to cut back on the level of financial resources dedicated to research (see also Jablecka, this volume). The financial squeeze was caused by the economic crises which accompany processes of political and economic transformation. The symptoms are the same in all countries: a significant decrease of GDP, a fall in production, inflation, and the rise in prices out-stripping the rise in incomes. The degree of decrease of funding for research can be only conjectured and it is almost impossible, on the basis of the available data, to establish relevant points of comparison across different countries. A catastrophic fall in incomes has occurred in Russia, the Ukraine, Belarus, and the Baltic states. The amount of funding for the Ukrainian Academy of Sciences fell in 1993 to roughly 10% of the funding available in 1990. In the Polish Academy of Sciences funding fell in 1994 to about 60% of 1991. In the Czech Academy of Sciences the real value of average monthly salaries decreased in 1994 to 86% of the real value of average salaries in 1989. However, since 1992 it has grown and in 1994 reached 110% of that found in 1992 (the nominal average salary has grown to 219% of that which pertained in 1989).

Looking at staff reduction more carefully, however, it becomes evident that it can hardly be explained away solely in terms of decreased financial support. Rather, it is essentially implicated with other intervening factors. It was clearly connected, for example, with the decrease – one can almost say the fall – of interest in the results of research on the part of both industry and state institutions. As a consequence, the shrinking state budget assumes even greater significance as it is the only serious source of research funding.

Further, staff reductions are implicated in the transition to competitive project funding for research (i.e., the grant system). This in turn implies a transition from a system of central distribution of financial means which largely disregards the quality of research projects, to a system which distributes finances to individual research centres on the basis of at least some minimal evaluation of their performance rather than their belonging to a specific ministry (see also Jablecka, this volume). The rapid pace of change to a system which now emphasises competitive funding of specifically-approved projects and which limits the distribution of institutional means has provoked fast reactions from many institutions that, historically, have been accustomed to having relatively stable financial resources. This change of the system has provided a significant impulse for encouraging staff mobility and often also for pruning out less efficient research staff.

Another reason for staff reduction in the Academy institutes was the dismantling of the artificially-privileged position of the Academies relative to the universities. As soon as Academy institutes opened themselves to the universities, the route for staff transfers from Academy institutes to the universities opened as well (see also Sarmir & Zajac, this volume).

These interleaving factors show that staff reduction in Academic institutes has not merely been a simple reaction to the reduction of financial means. Rather, we must understand staff reduction strategies as a key part of the process of dismantling the centralised model of science and research organisation that pertained under the conditions of command economies and authoritarian political systems. Most of the states of the ex-Soviet Union must grapple with the additional problem that their research systems were built as regional subsystems of a unified Soviet system of science, and thus do not sufficiently reflect the specific needs of these newly-autonomous countries. The Ukraine provides an illuminating instance. At the demise of the Soviet Union, it had the largest Academy amongst all the ex-Union states (Russia did not have its own Academy) which included 85 research centres and 94 experimental, construction and production centres. These organisations employed 69,000 people (including 16,000 researchers). Following political independence, the Ukraine must deal with the considerable discrepancy between its limited economic capacities and its inappropriately huge scientific potential. This discrepancy has led to an extreme reduction of funding for the Academy which by 1993 only offered 10% of the funding which had been available in 1990 (taking inflation into account).

Scientific research in the countries of Central and Eastern Europe was thus excessive in terms of both personnel and of the ratio of its cost to the GDP. The excessive number of research staff was a consequence of the old conception of a self-sufficient economy and of the limited possibilities for transfer of advanced technologies from the most developed countries. The requirement that economic development should be based on national research was an appendix to attempts at economic self-sufficiency. The conceptions of self-sufficiency and of extensiveness in research projected themselves onto the R&D structure resulting in its relative isolation from research in economically developed democracies, and its one-sided orientation towards co-operation within COMECON. Last but not least, the excessiveness of research staff was also influenced by an ideology which artificially gave precedence to science and technology over market rules, to state alimentation over free entrepreneurship, and which guaranteed full employment even where this caused inefficiency.

Over-staffing was interconnected with other circumstances that concerned the concept and organisation of the research system itself. For example, it is typical of the centralised model of research organisation that 30%-40% of employees do not engage in research directly. Rather, such employees engage in the activities of transferring research results which, in market economies, are carried out as part of a firm's innovation activities. In the centralised model of resource distribution, the employment of more staff is preferred to the allocation of monies towards building expensive technological infrastructures to support a leaner research labour force. One of the unfortunate consequences of the old centralised model was that, in addition to employing many excellent scientists, research centres retained many mediocre and less efficient staff members. The resulting imbalance between a relatively high number of employees and a low level of technological and information facilities was undoubtedly one of the significant factors which provided the impetus for major staff reductions during the process of transformation.

It is not yet clear whether the present extent of staff reduction in post-communist research systems corresponds to the extent of those countries' decreased GDPs. There is, in all industrial countries, a historically-established correlation between the amount of GDP on the one hand, and the amount allocated to expenditure on science on the other. The command economy distorted this respected correlation between GDP and the development of science; research activity was granted more resources than a relatively low economic efficiency would normally allow under market conditions. However, it is impossible to subsidise the imbalance between an extensively-developed research sector and a low level of GDP for long. Still, it would appear that even today in some countries of Central and Eastern Europe expenditures on research as a percentage of their GDP are higher than those of the OECD countries with the same GDP. In other words, their ratio exceeds the above-mentioned correlation and this remains the case even when we take expected growth of those GDPs into account. This assumption accords with the findings of those transitional economics (e.g., in Poland and Hungary) which have undertaken an international evaluation of their research centres and which have concluded with the recommendation that staff numbers in those countries' institutes should be further reduced.

Over-staffing of research institutes was supported in the Soviet-style societies by an ideologically-motivated wish that the share of research in the whole economic potential of the country would be artificially increased. It is not difficult to find a certain theoretical (and historical) correlation between this wish and the argument which justified the linear models of technological development that used to abound in the economic conceptions of the most developed countries during the 1960s. In the countries of Central and Eastern Europe, however, these arguments became the constitutive elements of ideologies that justified the advantages of central planning of economic processes. To a large extent, they permeated into the self-evaluation of the scientific community, for they proved useful in supporting that community's claims on the state.

They also gave rise to the idea, still widely shared in Central and Eastern Europe, that any research institution has a positive value simply by virtue of its existence – regardless of its relationship to world science and to the real innovation process in a market economy. As long as such an abstract view of research institutions survives, almost any staff reduction strategies will continue to be perceived as undesirable and harmful both to science and society. The vitality of this view is probably rooted in a self-deception: there is a certain illusion common amongst the rapidly-changing economies of Central and Eastern Europe that historically high levels of state support for research activity unquestionably count as an exceptional resource upon which these economies can build a sound future. However, as today's experience has shown us, under the economic conditions of open markets those kind of resources do not often correspond to the requirements of quality and thus are quickly devalued. It is highly likely that the reduction of research staff levels will continue until an appropriate level is reached which is compatible with economic growth. It is interesting to mention at this point that many institutes plan further reductions.

5. Two Strategies of Staff Reduction – Concluding Remarks

The goal of transforming research systems refers to more than their passive adjustment to the conditions of a rapidly changing social and economic context. Rather, the issue is to ensure that the foundations of a research system are actively put in place so that R&D activity may in principle correspond to the demands of the coming era. As is evident from the case studies, all the observed institutes have actively tried to cope – each in their own way – with a new situation which has radically decreased their research funding at a time when the societal prestige of science is diminishing and when industry and the state take little interest in the results of their research.

In this situation, institutes exhibit two tendencies towards meeting the demands of a changed social environment. On the one hand, there has been an effort to maximise their extant research potential; to save as many research posts as possible; to retain the continuity of their research orientation and research teams; and, as far as possible, to protect their existing organisational and institutional structures. On the other hand, we find a second tendency. There has been an intense attempt towards deeper integration into global scientific networks (see also Mirskaya, this volume). Researchers in Central and Eastern Europe increasingly accept the standards and norms of the international scientific community and reshape their own research work according to those standards. They are also starting to recognise that their native R&D systems are massively over-sized as compared with their OECD counterparts, and that such giantism is incompatible with the future health of science and research in post-communist societies.

Both approaches are legitimate and relevant in the new situations facing the transitional economies. The trouble, however, is that these two approaches are not always compatible. Given the rapid pace of change and the complex conditions which accompany the social transformation of a whole society, they can easily come into conflict with each other. Besides, choosing between them has never been a matter over which institutes have had a completely free hand. The actual choices made have mainly depended on the extent to which reforms in a given society have been radical or incremental; on the depth of the economic crisis in each country; and on the extent to which research funding has been reduced across different contexts. Depending on their immediate circumstances, different institutes have emphasised either the first or the second approach. Similarly, it is feasible for us to distinguish within our case studies two different strategies which institutes have used to protect their research potential.

In the first strategy, the effort to retain staff rather than embrace staff reductions was most in evidence. Despite funding cuts, the aim was to ride out the financial crisis by shrinking the means necessary for the successful conduct of research. This was a sort of defensive "survival strategy" of the extant research institutions and its staff members. It found expression, on the one hand, in the reduction of research investment, tangible costs and overheads and, on the other hand, in the distribution of finances more or less equally amongst researchers in the form of salaries. Where research institutes have opted for this strategy, structural changes have progressed slowly; the evaluation of research teams has tended to be neglected; and the long-term conception of reform has not been seriously taken on board. The energy of the research administration has tended to be wasted on unrelenting attempts at short-term survival whilst the institute waits for governmental institutions to make the more fundamental policy changes to improve (or restore) the conditions for research.

Institutes which have adopted the defensive strategy as a priority have opposed on-going staff reduction by all means at their disposal. Their tactics include sending their employees on unpaid sabbaticals, shortening the working week, postponing the payment of salaries or offering reduced salaries. Since the incomes of researchers in some countries are lower than the national average, these staff members have tended to take on second jobs – sometimes in areas close to their research activity but more usually in completely different areas. For example, 36% of researchers in the Russian institute of economics have a permanent second job while 43% occasionally take on a second job.

The explanation for why the defensive strategy is sometimes the primary response exhibited by institutes largely lies out with the institute itself and refers to the current difficulties and vague prospects of social transformation. In those societies where funding cuts were exceptionally great and sudden, and where scientists could only, with great difficulty, find jobs outside research because of the high rate of unemployment, the defensive strategy appeared as the most natural reaction. The lack of a clear and attractive vision of social transformation point in the same direction. Some of the studies provide evidence that at the institutes which hold a negative attitude towards staff reduction (e.g., at the institutes of the Russian Academy of Sciences), a half to two thirds of employees did not have a clear concept of the future of research and expressed disagreement not only with the current conception of reform in science, but with the conception of social transformation as a whole. When researchers fail to possess a vision of how to overcome an immediate crisis in research and a larger crisis in their society as a whole, the turn towards the defensive survival-of-science strategy is quite understandable.

The second strategy, in contrast, concentrates on improving the quality of research and on implementing selective staff reductions, which is understood as a necessary means to ensure the growth of productivity, efficiency and quality of research activity. In such cases, the efforts of the institutes' administration and top researchers are directed towards supporting internationally recognised research teams at the expense of mediocre and less efficient teams.[6]

The most important means towards selectively reducing staff numbers – those of research teams as well as whole institutes – is to engage in objective evaluations of efficiency. This approach was pursued most especially in the

6 The above mentioned strategies of staff reduction must be understood, nevertheless, as analytical simplifications. In the real process of staff reduction of individual institutes and in the opinions of respondents they appeared rather as tendencies which often interpenetrated each other. They apparently reflected the objective contradictory character of transformation and staff reduction. Nevertheless, in the explanation of an actual course of staff reduction it is methodologically useful to follow the conceptual clarity of both ideal types.

Czech Republic, Poland, Hungary and the Slovak Republic. Where it had taken place, it became the most efficient stimulus for subsequent transformation steps. For instance, in the observed three institutes of the Czech Republic comparatively objective quantifiable criteria of evaluation were implemented. Although many researchers had initially perceived the radical application of these criteria as a shock, since they had not encountered any such evaluation procedures previously, it quickly became clear to them that this was a key step towards a successful transformation. In all three analysed institutes, staff evaluations began as early as 1990, were repeated in 1991 and 1992, and in 1993 were applied with radical consequences for the staff. The basic process of staff change ended only in 1994 when, in all three inspected institutes, the complex evaluation of university-educated staff took place. These examinations have finally stabilised the jobs of researchers to a certain extent. Examinations were conducted according to the unified principles valid for all the Academy and are set to become in the future – at five-yearly intervals, at maximum – a regular part of the evaluation of staff's performance.

A major step in the same direction was established in 1993 with the system of independent evaluation of the institutes of the Czech Academy of Sciences. Its current expression is the first international evaluation of the quality and efficiency of institutes' research. For the first time in the observed institutes (and within the Academy in general), an evaluation is under way which is, as far as its principles and methods are concerned, on a par with the norms and customs of the OECD countries. As the case studies make evident, the employees of the observed institutes experience it as a significant change, for it is crucial to supporting their competitive position in an environment which increasingly differentiates its support for particular institutes.

IDENTITY PROBLEMS: BASIC OR APPLIED RESEARCH?

GYÖRGY DARVAS

Institute for Research Organization
Hungarian Academy of Sciences
1361 Budapest

1. Introduction

Most of the institutes belonging to the academies of science in Eastern and Central Europe were originally founded to perform basic research and were financed from state budgets for that purpose. The share of R&D expenditure from each national income – and within that, the proportion allocated to fundamental research – far exceeded the rate that would have been justified given the relatively modest wealth of these developing economies. During the '80s, most of these states became indebted and their national budgets were in negative balance. Thus, the prevailing high rate of government subsidy could not be maintained. According to both general policy and an enforced ideology, science was billed to become a productive force. If it could not do so by itself, then state policy should facilitate that aim. While applied research more or less continued to receive support, institutes engaged in basic research were now expected to find external sources of income from contract work for applied and even development tasks – although the specific shape of those expectations differed between countries. This policy shift was first evident in Hungary, then Poland, and was later followed by most of the East and Central European countries.

The dynamics of economic and political transformation since the early '90s have not improved the financial health of these states. In turn, the possibilities for government financing of research have become even more perilous (see Jablecka, in this volume). At the same time, government control over the academies, and academies' control over their individual institutes, has weakened. The freedom of institutes to choose the type of research they engage in has increased. The general question to be addressed in this chapter is whether the type of research pursued in the academies' institutes has changed or remained constant during the period of transformation. And, if it has changed, in which direction is it now heading?

Beyond a mere description of what has happened to the institutes, we are interested here in the causes which underpin either the changes in, or the stabil-

R. Mayntz et al. (eds.), East European Academies in Transition, 49–61.

ity of, the types of research being undertaken. We distinguish between determinants external to each respective institute, and determinants internal to it. External determinants consist of opportunities which have opened up during the period of transformation, as well as pressures and restrictions which propel an institute in a particular direction. Especially relevant here are changes in research policy, and changes in external demand for research results in industry and other societal sectors. Internal determinants consist of the strategies by which different levels of actors within an institute react to external opportunities and pressures. Here we can distinguish between the strategies of the institute as a whole (as represented by its directors) and the more local strategies of research groups or individual researchers. Both kinds of strategy may conflict with each other. We can detect different patterns in the trajectories different institutes follow with regard to their research orientation, depending on the particular external conditions they face and the particular responsive strategies they adopt to manage those conditions.

Many aspects of these dynamics have varied greatly between institutes. However, our 37 case studies across 12 countries provide enough coverage to encourage us to formulate certain generalisations. Although the context of each institute differs such that no two are in the same situation, some common tendencies are very apparent. This chapter will concentrate on these commonalities at the risk of neglecting a detailed analysis of some local phenomena – albeit that they are important at that particular level. Analytically, different patterns of constraints and options with regard to applied or basic research can be detected.

2. New Emphasis on Basic Research

When the centrally planned political regime was overthrown, one immediate result was that direct political pressure in the form of political directives on the different levels of research actors – the academy as a whole, its institutes, research groups and individual researchers – was greatly weakened. The new political regime granted research actors more or less similar kinds of autonomy which are common in Western countries. This included that immediate government influence through shaping the financial conditions to engage in applied research – a feature of the old order – temporarily ceased.

Under these circumstances, most institutes of the academies altered their research orientation towards the pursuit of basic research. It is not self-evident why this should have been so. While political change had certainly created a new opportunity to depart from applied research – the most prevalent orientation at that time – institutes were not forced to change direction. They might simply have stuck to doing applied research. However, the prestige attached to engaging in basic scientific research is pertinent to our understanding of institutes' change of direction. In the normative self-conception of researchers world-wide, basic research enjoys the highest prestige. When this self-conception is internalised by researchers, it becomes an internal determinant of an institute's preferred research strategy. An example of such an attitude was voiced by a researcher in the Romanian institute of physical chemistry: "We finally got rid of the pressure [to do] applied research, and we'll fight for the maintenance of basic research." Nevertheless, the origin of this self-conception is external to any particular institute; it springs from the global academic culture. Thus, even researchers who do not themselves share the view that basic research is necessarily more prestigious, nevertheless must accommodate the fact that the global academic culture to which they belong does rank basic research very highly. They work and communicate in scientific communities, both internationally and in the narrower circles of their immediate colleagues, where they are evaluated according to their basic research achievements, and this reputation is a crucial determinant of their career chances. The cultural superiority of basic research seems to be especially strong in Russia, Belarus, and the Ukraine. In these countries, during the second phase described in the next section, researchers felt their engagement in applied research to be work beneath their abilities; they regarded it as both a shame and a temporary necessity, justifiable only in order to secure the survival of their work places.

One additional reinforcement of this culture must be mentioned. In many of these countries, measures to increase the efficiency of the research system have been put in place, including evaluation procedures. In this respect scientometric indicators of the research performance of individuals and institutes have frequently been introduced. But the indicators typical of this kind of assessment exercise, such as publication rates, are biased in favour of basic research. Basic researchers publish more papers in contrast to applied researchers, whose most important results are not publications but patents or technical prototypes. Consequently, the recent introduction of scientometrics into research policy has emphasised the traditional academic culture dominated by basic research. For instance, in the Belarussian institute of engineering the directors increased the importance of scientific publications in their estimates of their researchers' activities, which was viewed by them as stimulating basic research.

Of course, the cultural superiority of basic research was also used by institutes to further their more general interest in augmenting their autonomy. Whenever the autonomy of science is stressed both by researchers and by political actors, the accent is usually placed on the autonomy of basic, as opposed to applied, research. Thus, presenting one's work as basic research is an implicit safeguard against intervention from research policy-makers into an institute's work and decision-making. It should be kept in mind that such interventions usually had not consisted of specific political commands which research actors had had to obey. Rather, political actors shaped an institute's mode of financing in such a way that only applied research could secure an adequate level of funding. Since most institutes endured such interventions throughout the '80s, they were motivated to pursue basic research as one kind of defence against this kind of external pressure. It was well known that in the Western countries basic research activities often enjoy funds from the state without any further stipulations of content or goals of research.

This reasoning was also evident at the level of the academy leadership. It was not only the institutes but also the academy as a whole which had been subject to political influence. Thus, in all countries the leaders of the academies declared that a return to basic research was to be an essential element of the academy's new corporate identity (see also Simeonova, in this volume). For the institutes, this meant that they had to comply with the general policy of their academy even in situations where, left to themselves, they would have tended to remain oriented to applied research. The most extreme example of this is found in the three Romanian cases. All three institutes had belonged to the branch sector prior to the period of transformation and were strongly committed to doing applied research. When the Academy of Romania was reorganised, a number of branch institutes wanted to return back to the academy to which they had belonged before. This wish was accepted by the academy – provided the institutes were willing to focus on basic research. The institute of theoretical informatics, for instance, had to officially declare its change in research direction. The new statute of the Academy of Romania contains the formal requirement that its member institutes must orient themselves to basic research.

Finally, government actors also tended to promote basic research. In some countries, such as Hungary, this echoes back to the activities of the '80s. At that time, basic research accounted for an extremely low proportion within the national R&D budget and governments sooner or later felt obliged to throw a lifebelt to their basic researchers. This was done by establishing special funds for promoting basic research. Following the period of transformation these agencies fitted into the general model of similar Western institutions. Thus, throughout the region, national funds for the promotion of basic research were set up. Some of the institutional finances which had previously been assigned to the academy were withdrawn and then redistributed in the form of grants.

For some institutes, initial hopes of a shift towards basic research or of improving the prospects of maintaining such research, have been fulfilled. These institutes have experienced a stable correspondence between the internal and external determinants of this type of research. Not only are their researchers devoted to basic research, but their academy as well as available national and international funds direct them towards this very activity. One example is the Russian institute of theoretical physics, a world-renowned centre of excellence in that scientific field. The main objective of the institute's activities, both historically and at present, is, as its representatives declare, the "general development of scientific knowledge independently of its concrete practical application." The amount of basic research done as a proportion of the institute's total activity ranges from 70% to 100%, according to the estimates of the experts we consulted there. All the researchers based in this institute are currently engaged in basic research, considering their studies to be both scientifically promising (95%) and intrinsically attractive intellectual work (95%). Many of the researchers in this institute note that the actual topics being studied have changed during transformation. However, they also note that these changes are related, in the first instance, to global trends in science, and not to political pressures. Another example is the institute of physics in the Czech Republic, whose directors proudly state that "the institute is currently performing neither mission-oriented basic research nor applied research" but pure curiosity-oriented basic research.

3. Applied Research as a Survival Strategy

Most of the institutes, however, did not and could not maintain an orientation towards basic research. For those institutes which in any case wanted to do applied research, it was opportune that the increasing scarcity of resources forced all institutes in the direction of applied research. As a consequence of the economic and fiscal crises now facing these countries, governments massively reduced their institutional funding. This forced institutes to earn money – and they hoped to acquire it by forging contracts with industry and with other users of their research results. Many political actors hoped that in the long run institutes would learn to become responsible for their own survival, to adapt their work to the needs of societal users, and to largely finance themselves.

This maintenance of or return to applied research was often not the consequence of a deliberate strategy taken by any of the institutes as a whole, as had been the case with the preference for doing basic research. Wherever that happened, both institute leaders and the majority of their researchers had acted in concert as a single corporate actor. Concentrating on applied research, by contrast, was characterised more by individual actors coping with the new conditions they faced. The actual choices made by institutes were simply the result of a plethora of fragmented decisions and strategies made by individual researchers and research groups. These "micro" decisions and strategies were, in turn, the actors' responses to the often unclear "signals" they received from research policy, Western sources, the market for research contracts, the funding agencies, the international scientific community, and what they observed other researchers doing in comparable situations. Such signals were the basis for individual strategies which brought about the "macro" changes of an institute's overarching research leanings.

Concentrating on applied research was often involuntarily on the level of individual researchers, given that the majority of them intrinsically preferred doing basic to applied research. Only in a very few institutes was applied research genuinely preferred. In some institutes, as in the Slovak Republic institute of physics, there were a significant number of researchers who combined an orientation to basic research with an openness towards the potential application of their research results: "There exists a group of researchers who demand an exclusive orientation towards basic research together with complete freedom to choose their orientation, and researchers who consider it necessary to conduct both basic research and, to a rational extent, applied research as well."

Usually, however, having to do applied research was clearly regretted. For instance, researchers and directors of the Russian institute of biology spoke with sorrow about the shift towards applied research. Many researchers there had to make a very painful decision: either to stick to basic research and so give up the last chance for their families to live decently, or to earn money through doing applied research. The same thing happened in the Ukrainian institute of informatics. Although a certain historical orientation to applied research had already existed in this institute, researchers noted that, during the nineties, there was a pronounced imbalance in this research direction so that, even in this institute, it was considered as a disliked obligation. As one director expressed it: "Life forces us to pick up applied research tasks." A manager from the Belarussian institute of physics said of applied developments in that institute: "Our developments are used somewhere, someone is interested in them, and we do our best to implement them. Very often it is not as a result of our wish to change our basic research topics but of the need to survive at this hard time." Another example was the already mentioned Russian institute of biology. Researchers there considered it to be a waste of their time to manage their own research in the sense of marketing it or of winning contracts – even where they were successful in earning money: "The Russian market for scientific products in applied R&D is being formed. ... You must learn to orient to the machinery and the technology markets, adjust to them, study the clients and competitors. It can be done by specialists with the appropriate [skill] profile. Researchers are naturally professional, they can master science management easily enough. But it is a very sudden change. Enthusiasts of scientific research think it impossible to change their lives in this way. And even if they do change, it can be considered a loss for science." A very pessimistic assessment came from the head of a department of the Ukrainian institute of chemistry: "Basic research in this department – there's no doubt about it – dies. We did not break down completely because we earn additional money on our own initiative. We work on topics we should not be concerned with. I do it to maintain the department, but our basic research shrinks catastrophically."

Some institutes have been quite successful in earning money through applied research. But the kind of applied research for which they were able to win contracts was often of a very short-term character and of very little scientific interest. In many cases, research in the proper sense was not required – the work amounted only to testing or consulting etc. Increased involvement in such activities leads to a gradual degradation of genuine scientific creativity. Such applied work simply does not pose adequate intellectual challenges. Thus, securing the institute's survival by means of doing applied research implied the danger of losing one's scientific competence.

With this unwelcome prospect, the institutes, in several cases, were at least able to maintain some basic research activities. They managed to finance their genuine research interests by applied work. Thus, from the Ukrainian institute of informatics it was reported that "we [managed] to maintain the extent of basic research" although applied work was taken up "because with its help we could keep up our traditional research topics." In the Romanian institute of physical chemistry, its directors and most of its researchers agreed that applied research was necessary to finance basic research. The additional money was used for the institute's technical infrastructure, its library, and the travel costs of sending researchers to conferences. Moreover, applied work was supposed not to interfere with basic research. In the Ukrainian institute of physics, small innovative spin-off firms were founded part of whose earnings were to be used to finance basic research.

Earning additional money by applied research presupposes, of course, that there is a demand for such work. Two main potential users of the research results of academy institutes may be distinguished: applied research institutes, usually former branch institutes, and industrial firms. Both kinds of potential users had their own grave problems, however. As a consequence, many academy institutes turning to applied research quickly learnt that their efforts were in vain. The demand for what they had to offer was small and became even smaller over time.

The applied research institutes disappeared from the scene in almost all countries. Neither government departments nor private industry maintain more than a few applied research institutes. The lack of this very important sphere of institutes widens a deep gap between the basic research institutes and their environment. The problem can be illustrated by the case studies from Belarus. The breakdown of the USSR and the resulting economic conversion and subsequent economic crisis resulted in a weakening and, occasionally, a complete disruption of relations between Belarussian academy institutes and applied research institutes in both Russia and the Ukraine. In 1994 the Belarussian institute of physics had as few as three contracts with research institutes from the CIS countries, while in 1990 their number was ten times larger. Contracts with Belarussian institutes of applied research similarly weakened. The Bulgarian institute of computer and communication systems reported: "Relations with industrial institutes are as a whole in crisis because of the catastrophic state of industrial science ... Nowadays, when branch institutes, with few exceptions, do not exist, it is hard to speak of significant relations with them." The Polish institute of physics co-operated with applied research institutes mostly in the form of commissions from these institutes to conduct experiments or tests in specialised labs rather than in joint research projects, as had been the case in earlier times. Thus, even if contacts still existed, they were no longer as intense or as fruitful as before.

The industries in the countries under consideration showed very low levels of interest in implementing the achievements of domestically-produced basic research – although this trend was also differentiated by which kind of industry was most relevant to the institutes. For example, for almost all the investigated academy institutes in Russia, Belarus the Ukraine, and also in the Baltic countries their main client had been the defence industry of the former USSR. With the collapse of the union, the defence industry ceased to place orders with those academy institutes.

In other countries, the defence industry has not historically enjoyed that level of importance for the institutes. However, civilian industries have also been in deep crisis in all the countries under discussion. One may think that the adoption of technological innovations based on research results from academy institutes might be a way out of this crisis. But this strategy was almost never chosen. In the Czech Republic, for example, very few enterprises specialise in advanced technologies – and these are the only kind of companies which are interested in the results of scientific research. Most companies are oriented towards a quick return on their investment and towards short-term profit. A strategy of science-based technological innovation demands that a firm adopt a long-term perspective; a strategy that is simply unaffordable when the firm is struggling for its day-to-day survival. This is evident in the case of state-owned enterprises. As regards the privatised part of industry, the lack of domestic capital has meant that significant parts of those industries have been purchased by foreign investors. The attitude of such investors has been to avoid state-of-the-art high-tech production in the local regions they have bought into. Their purchasing decisions were partly premised on acquiring new markets, but also partly on the availability of a relatively cheap labour force. In terms of the level of production technologies deployed locally, branch sites do not take precedence over the parent companies' production methods, where the newest technologies and newest products are concentrated. In the few cases where new technological developments have been introduced, they have not been based on locally-devised applied research. Thus e.g., the Estonian institute of chemical physics has had no research contracts with industry in the past five years. Similarly, a respondent from the Estonian institute of chemistry stated: "..Estonian large-scale enterprises that used to be our biggest customers as regards the contractual jobs, do not order anything much." Thus, almost all institutes have had to come to terms with the reality that closer contacts with industry may, at best, only emerge at some future date.

An additional reason for the lack of interest of these countries' industries in the applied research of their academy institutes was the former COCOM embargo, and its abolishment by the eve of the demise of the socialist regime. As long as this embargo had been in place, many academy institutes had been engaged in 'reinventing' technological applications already used in the rest of the world, although they were not available to local industry. Once the embargo lifted, several institutes thus lost their main justification for doing applied research work. Indeed, a whole raft of similar problems emerged from the collapse of the COMECON. For example, applied research activities in the Bulgarian institute of nuclear research had been oriented towards the priority areas set up in the framework of the COMECON; after its dissolution, the organisational, as well as the financial and economic, framework vanished which had shaped both the prevailing problematics, and the intensity, of applied studies.

When it became apparent that there was no demand for applied research from academy institutes, research policy might have tried to stimulate such demand by, for instance, offering firms financial incentives to award research contracts. But this did not happen anywhere. For example, in Bulgaria, representatives of institutes complained about the state's refusal to function as a patron or facilitator. In Poland, the lack of an intermediate transfer organisation was mourned by researchers of the institute of biology. Researchers there had the impression that they might be useful to the pharmaceutical industry. One interviewee told that "in Poland there is no established mechanism for translating the accomplishments and specific proposals of research institutes to improvements in industry [at the intermediate level]. This is all the more necessary because basic research results cannot be applied directly in practice on an industrial scale. No specialists for this intermediate stage are being trained." Some institutes even began to miss the former system of central political planning which prescribed certain institute-user relations in long-term programmes.

With the few remaining applied research institutes – as well as firms – being uninterested in the applied research of academy institutes, the latter's positions became rather precarious. Lacking customers for their research results reduced their justification for existing in the eyes of the relevant political actors. Had the switch to applied research been successful, the institutes would not only have earned the additional money they needed, but they would have also legitimised themselves according to the goals of research policy. Being mainly unsuccessful, however, the institutes not only had to find other sources for the money they needed, but they also had to find a new legitimation for their existence.

4. What Remains is Basic Research

Since applied research was futile because there was simply no demand for it, the share of basic research increased automatically – as was reported in the Bulgarian case studies: "Increase of the amount of fundamental research is only relative in comparison to other decreasing activities." When applied research decreases, the relative share of basic research increases even where the total amount of research undertaken decreases as a whole. But beyond this single tendency, basic research has been emphasized in many academy institutes as a consequence of the interplay of several factors.

First, institutes and researchers had to digest the experience of unsuccessful applied research and, as already mentioned, to legitimise their new situation to the academy leadership and to political actors. If they could not show any successes with regard to applied research, they could not present themselves as doing this type of research. What else, then, could they say they were doing? They could only declare their turn to applied research as an error and assert that, from now on, they would firmly adhere to conducting basic research. This kind of legitimising 'statement' could be, and was, a statement of conviction since this type of research was, indeed, what the overwhelming majority of institutes and researchers had sincerely chosen as their genuine mission after the abolition of the centrally planned regime, whereas the switch to applied research had obviously been involuntarily in most cases.

The reorientation to basic research was supported by the ongoing policy of the academies to present themselves as institutions pursuing this type of research. All the academies have continued to expect their institutes to concentrate primarily on basic research. This expectation has been especially forceful in cases where the intra-academy distribution of institutional funds has depended on an institute's adherence to such a profile. For instance, in June 1995 the Belarussian institute of biology faced the verdict of the authoritative scientific commission of the academy which had evaluated the institute's performance over the previous five years. The commission's recommendation was "to enhance basic research works in the priority research trends, rendering special attention to mastering and using new advanced methods and approaches."

Another factor which has driven academy institutes into basic research has been the national and international funds which predominantly support basic research. A representative of the Bulgarian institute of computer and communication systems stated quite succinctly: "... in the absence of budgetary means – only wages are guaranteed by the budget – as well as of serious incomes from contracts with industry, with the backward technologies at hand, the development of theoretical research raises doubts, despite the presence of highly qualified researchers...The absurd fact that the library of the institute is closed and the lack of any means for subscription to scientific journals seems to make these doubts justified. After private firms have been closed, the only source of additional income remain those from competitions of EC, PHARE, TEMPUS and the National Fund for Scientific Research (NFSR)." Similarly, experts in the Russian institute of radio engineering and electronics highly estimated the importance of the Russian Fund for Basic Research, considering it to be a promising funding source for basic research developments. The institute has closely interacted with this funding body together with other international funding sources. The transformation of the Belarussian institute of physics was supported to a considerable extent by the Belarussian Basic Research Foundation. In particular, according to the experts, it is the Basic Research Foundation that helped to stop the prevailing development of applied research at the institute.

With all these external and internal factors in play, it is still important to carefully consider to what extent the orientation towards basic research has really occurred, and to what extent it has been only rhetorically declared and remains, at best, a vague hope for the future. The deterioration of research conditions has stopped many research activities completely. Thus, too often, neither applied nor basic research has been undertaken to any significant extent.

Considering the desolate financial situation of many institutes, it is not surprising that their actual or declared orientation to basic research is also a matter of political rhetoric. This type of research is useful for justifying claims that each state must take at least some financial responsibility for the continued work of its institutes. The most pronounced arguments of this kind have been put forward in Russia.

Directors and researchers of the Russian institute of theoretical physics see the main reasons for the crisis in academic science as being caused by insufficient state financing and lack of governmental support; by the low level of public demand for the results of basic research; and by the extreme bureaucratisation of the Russian Academy of Science. In the experts' opinion, the entrepreneurial sectors will not, in the near future, finance basic research. Therefore, the main form of financing basic research can only stem from government. Individual experts have stated that basic science, which will ensure the country's future economic development, "must be" financed by the state at a high level. Any financial independence from the budget "is out of question: this is the prestige of the country and the state". Another expert reported that the "state is obliged" to provide basic research: "All the rest may be merely supplementary, and not vice versa, as it is now." The creative potential for enhancement of basic research is available, but to realise it, "it is necessary to change the science policy of the state." It is interesting that this opinion is also held in applied research-oriented institutes such as the Russian institute of radio engineering and electronics. As one of the experts there states: "Financing of basic research is one of the state's prerogatives, it is what makes the state, a state." According to the same opinion an entrepreneurial sector will not, and should not, finance basic research: "This is not its function, it contradicts the aims of the sector", especially when it is just emerging and is, as yet, only oriented to short-term profit. Certain experts make categorical statements by referring to foreign examples: "All over the world, basic research is primarily financed by the government." They are convinced that basic research institutes cannot in principle be independent from state budgets. Alternative sources of financing should only be secondary and supplementary. In a situation where such alternative sources are deemed to carry the major funding burden, these experts maintain that basic research can only suffer. The experts are sure that "without breakthroughs in basic research it is impossible to achieve technical and technological progress and, consequently, economic and political superiority (sic!) in the world. It costs a lot of money, which the state's leaders must understand." Thus, in the opinion of these interviewees, the state has more than just a moral responsibility for financing basic research; state funding is seen as a functional necessity if the country is to prosper again.[1]

5. Outcomes

To ask whether basic research has deposed applied research or vice versa, is to pose the wrong question. In the countries we investigated, both types of research have lost ground in the period of transformation. Neither kind of research orientation could convince either the political or the business spheres of the importance of research, or of the institutes' partnership in the battle for the political and economic transformation of their respective countries.

More specifically, one may distinguish analytically four types of situations the institutes have faced. On the one hand, an institute's genuine preference may be either for basic or for applied research. On the other hand, which type of research an institute actually does is a different issue entirely. With this simple typology, the described constraints and options can be summarised as follows:

- In the last decade before the transformation, most institutes wanted to do basic research but were compelled to turn their hand to applied research; a minority preferred doing applied research.
- The first phase after the break-up of the old regime was characterised by an emphasis on basic research. Many research groups and individual researchers could bring their actual engagement in research closer to their wants. During this phase, those institutes which preferred applied research were confronted with a tension between both research orientations. The Ukrainian institute of informatics provides just such an example: its major research activities had always been applied, and that was what most researchers and all directors wanted to continue, but they had to give up many of these activities in favour of basic research.
- Very soon economic and political realities had to be faced by the institutes. Many were forced into applied research. For them, their preferences and their actual work became incongruent, whereas those institutes which had been driven in the first phase into a situation of incongruence were now relieved from this tension. The incongruence of doing – but not

1 One may detect a thread running throughout these statements of the survival of overemphasised expectations addressed to a paternalistic state.

wanting to do – applied research was doubly frustrating because, in most cases, engagement in applied research failed as a survival strategy.

- The result of this breakdown of applied research is that many institutes by now are unable to earn the money necessary to maintain their research conditions. This means that researchers are not compelled to do applied research they do not want to do. But they are also often unable to do the work they do want to do. Only a few institutes have been relatively successful in preserving their research conditions such that they realistically continue to do basic research.

In sum, the present situation remains highly unsatisfactory for most institutes. The prevalent rhetoric of doing basic research is often nothing more than a white lie and a self-deception.

WHAT DETERMINED AN INSTITUTE'S FATE IN TRANSFORMATION?

UWE SCHIMANK

Fernuniversität Hagen
Fachbereich ESGW, Soziologie
D-58084 Hagen

1. Introduction

Institutes of the national academies of science of post-socialist countries in Central and Eastern Europe have experienced quite different fates in the period of transformation which is still in progress. These societies' transformations have consisted not only of far-reaching political, economic, and cultural changes but have included massive changes in the national research systems (see the national reports in Mayntz, Schimank & Weingart 1995). For scientific research, these transformations have resulted in a coincidence of opportunities and trouble (Schimank 1995). On the one hand, political rebuilding has brought about opportunities for a greater self-regulation of research, which has been demanded by many researchers for a long time; on the other, an appropriate institutional rebuilding of the research system has been greatly restricted by sharply reduced resources as a consequence of the deep economic crisis in these countries. Thus, over time, all academy institutes, like the research system as a whole, have been confronted with shrinking opportunities and growing trouble.

Most academy institutes have suffered significantly from this in various respects. But the amount of trouble has varied considerably not only from country to country,[1] but also from institute to institute; and the institutes have differed substantially in their ability to cope with this trouble. As a consequence of both differences, some institutes have been relatively successful in maintaining or even improving their research conditions while others have more or less failed in this respect.

But what have been the determinants of the amount of trouble an institute has had to face and of its capacity to cope with this trouble? Besides many idiosyncratic factors which certainly have been present in each case, three factors in particular have been pointed out repeatedly as strong universal determinants of

1 Variance between countries will not be taken into consideration in this analysis.

R. Mayntz et al. (eds.), East European Academies in Transition, 63–78.

an institute's relative success or failure: scientific quality, type of research, and academic field. For each of these determinants, a general supposition can be formulated which expresses a widely held view:

1. Institutes which, according to intra-scientific criteria, had been conducting high-quality research before transformation have been more successful during the transformation than low-quality institutes.
2. Institutes which could continue to conduct applied research or have been able to quickly shift to applied research have been more successful during the transformation than institutes which have remained in curiosity-oriented basic research.
3. Institutes in the natural sciences, including the medical, the agricultural, and the engineering sciences, have been more successful during the transformation than institutes in the social sciences and humanities.

Without spelling out the reasons given for these suppositions at this point, it can be deduced that a high-quality applied natural science institute should have had the best chances of success, whereas a low-quality curiosity-oriented institute in the social sciences or humanities should have had the lowest chances of success.

This chapter will examine how much truth there is in these suppositions. How much of the variance between the cases can be explained by each one or by a combination of these three factors, and do they indeed work in the assumed causal direction? But before this question can be approached, we must clarify the indicators by which the relative success or failure of an institute can be assessed.

2. Indicators of Relative Success or Failure

How can the relative success or failure of an organization be assessed? This is a very difficult question in general, and even more difficult in the particular cases of academy institutes during the ongoing transformation. To begin with, any estimation of success or failure is bound to a specific time horizon. Success is not for ever, but may turn any time, and sometimes quite abruptly and unexpectedly, into failure; and failure, if it does not go to the extreme of an organization's ultimate extinction, may turn into success. As long as we study only an organization's past, this fact can be taken into account. But very often, and certainly with respect to our cases, we are interested in an organization's future prospects. The transformation of post-socialist countries is not finished yet but will continue for several years, and the same is true for these countries' research systems. Whether an academy institute may ultimately survive all that is still to happen, and to what degree this institute will be able to maintain or even improve its research conditions, is an open question even for those institutes which to the present day have been comparatively successful. Accordingly, all assessments of an institute's relative success or failure are strictly for the present; and, taking into account the turbulent dynamics of these countries' transformations, any extrapolation into the future is even more risky than in general.

But this is not the only difficulty of evaluation. Which criteria are adequate to measure success or failure? There is only one extreme situation which is relatively easy to categorize, and that is if an organization has definitely ceased to exist and this has happened against its will. Even this apparently obvious case of complete failure may harbour certain ambiguities. As a corporate actor, an organization is a collectivity of individual members who may not achieve total consensus on its desirable fate. Whose opinion is the relevant one under such circumstances? The majority's, or the leadership's? This lack of consensus in the estimation of an organization's situation will also be a problem with regard to all other criteria which become relevant if an organization has survived. For instance, an organization may have reduced its personnel considerably. This might be assessed as relative failure by many members and especially by those ex-members who have had to leave. But the leadership might perceive this occurrence as relative success because the organization got rid of inefficient slack and is now better prepared for a difficult future. Thus, although both sides have used the same criterion, they have arrived at contrary views about the organization's success or failure.

If survival as the minimum criterion of organizational success is accomplished, as is the case for all academy institutes considered,[2] problems of assessment multiply because then a plurality of plausible subsequent criteria of relative success or failure come into view. To list just some of them relevant for an academy institute: Has the institute managed to maintain its size or even to grow, or has it suffered reduction of personnel? Has it been able to attract and recruit younger researchers, or have its research employees been steadily aging? Has the research infrastructure (laboratories, equipment, libraries) deteriorated or improved? Has the institute been able to maintain or improve its level of performance, be it with regard to publications or to the transfer of research results to extra-scientific users, or has performance declined? Have the institute's financial means obtained from institutional funding grown or decreased? Has it been able to acquire additional financial resources from project funding or research contracts, or has it had to make do with institutional funding? Has the institute maintained or enlarged its international contacts, or have they deteriorated? There are only a few cases in which all of these criteria point in the same direction. But where they do not, relative success with regard to one criterion must be weighted against relative failure with regard to another, without the benefit of a common denominator or an evident rank-order.

2 No cases of complete failure were analyzed because it would have been too difficult to find persons willing to narrate what happened, and to narrate it without a strong bias in one or the other direction.

All these problems of judgment do not allow firm conclusions to be drawn about the relative success or failure of an institute.[3] But some overall impressions with regard to the aforementioned criteria can be summarized before setting out to discuss in detail the supposed determinants of an institute's relative success or failure (for more detailed information about several of these aspects, see also Jablecka; Provaznik et al.; Nesvetailov; Mirskaya, all this volume):

- All of the institutes have had to suffer very serious financial reductions. Still, the variance of financial losses among institutes should not be overlooked.
- The institutes' ability to acquire additional financial means by research contracts or project funds has varied greatly. Although none has been able to compensate fully its losses of institutional funds, a few have been able to reach a financial situation which has allowed for some improvement in infrastructure and researchers' incomes. In other institutes research literally has ceased more or less completely because the infrastructure has not allowed it to continue any longer and researchers have been preoccupied with other ways of earning their living.
- In general, the institutes' infrastructure is outdated, often needs repairs which cannot be paid, or has even broken down completely. Especially harmful in this respect has been the institutes' inability to buy Western literature and journals. Thus, most researchers are still isolated from international scientific communication, albeit for different reasons than before.
- Reduction of personnel has varied much among institutes. There are even some instances of growth. But in several cases reduction has been an explicit policy of the institute, thus cannot be counted as evidence of relative failure. Almost everywhere the reduction of researchers has been much less than the reduction of other personnel. Accordingly, most institutes claim to have been able to preserve their "scientific core".
- With very few exceptions, all of the institutes have suffered from the aging of their research personnel. The extent of this aging varies somewhat. But even those institutes with relatively young research staffs may be drifting toward a problem of the same magnitude as the institutes where the negative consequences of aging for research are already apparent.
- The scientific quality of the institutes' research is most difficult to assess. Taking publication rates, especially in refereed Western journals, as an in-

3 An additional problem will be only mentioned in passing. It consists in the fact that the reports of the case studies sometimes lack precise and reliable information about an institute's situation with respect to one or the other of these criteria; or the empirical indicators which are used are not comparable over cases. This well-known problem of any empirical research is an especially serious one for these case studies, not least because the institutes investigated had an inclination to gloss things over in their often very difficult situation.

dicator, we find that these rates went up during the first years of transformation. But these publications mainly reported findings from research done prior to transformation, and the sudden possibility of publishing in Western journals produced a quick increase in such publications. Wherever such data are available, it is clear that publication rates dropped in 1994, and this may reflect the decline of research conditions and scientific quality after 1990.

– The institutes' attractiveness to extra-scientific users of their research findings – the other important dimension of many institutes' performance – is in most cases rather low. In this respect, institutes have more or less failed.

Simplifying this summary even more, the conclusion is: More or less all of the institutes have failed in relative terms with respect to the preservation of their former level of institutional financing, provision of infrastructure, inflow of young researchers, and maintenance of research performance, although a considerable number of institutes have been relatively successful with respect to the acquisition of additional financial resources and the keeping of their personnel, especially their researchers. The overall picture emerging from this is that survival has been achieved, at least for the time being, but all further aspirations have been largely unfulfilled. In this respect, the selection of the cases studied seems to be quite representative. On the one hand, only a small number of academy institutes has actually been closed; on the other, there is not a single one among those studied which is really well off.

3. Supposed Determinants of Relative Success or Failure

It is clear that relative success or failure is mainly the result of forces outside of the respective institute, and beyond its influence. Yet to some, varying, extent, an institute may be able to influence its fate by its own activities, including strategic acts of coping.[4] We should bear this in mind as we now come to discuss the three supposed general determinants of an institute's relative success or failure mentioned above. All three determinants are properties of an institute and its activities: the scientific quality of its research as well as the type of research and the academic field. The supposed importance of all three determinants is related to changes brought about by societal transformation dynamics beyond the institute's control. The determinants differ with respect to possible coping activities on the part of the institute. Its academic field is something an institute cannot change. The same is true of its past scientific quality, which can only be enhanced by means of impression management. However, an institute

4 In a study of the institutes of the dissolved East German academy, Hans-Georg Wolf (1966) discusses this interplay of external forces and an institute's own activities very carefully.

may be able to promise or persuasively demonstrate quick improvements in scientific quality. The type of research done by an institute, finally, is something it is able to adapt strategically.

3.1 SCIENTIFIC QUALITY

It seems plausible that in times of trouble those organizations which have performed the best should have the best chance of survival. Moreover, this is not just an empirical fact one expects to find; it is often perceived as a desirable and just result. From the point of view of research policy, it is logical to make a virtue out of necessity. If reducing resources is unavoidable, these losses should be distributed among academy institutes in a way which eliminates mediocre research. This desired outcome also conforms with the meritocratic idea of fairness.

A number of cases studied do more or less confirm the supposed correlation of scientific quality and relative success of an institute. The clearest case, in this respect, seems to be the Russian institute of theoretical physics. It was an international "center of excellence" in its field before transformation. Its personnel has not decreased but increased, which is a very rare exception. Moreover, the institute has been able to recruit young researchers. It has managed to strengthen its international networks and participates in several collaborative research projects with Western partners. Most of its researchers have acquired international grants, so that about half of the institute's financial resources come from Western sources. Although many researchers have taken up other work besides research to earn some more money for themselves, this work often has been connected with their research projects. In addition, from the point of view of researchers, the institute's infrastructure has at least been preserved at a satisfactory level. Thus, although the institute has not been able to maintain the formerly very high quality level of its research, research performance has remained quite high. This relative success of the institute is attributed mainly to its international scientific standing.

A more ambiguous case is the Latvian institute of polymer mechanics. Before transformation, it had been conducting high-quality research for a long time. It had been the center of a specific scientific school and had received a very positive evaluation from an international commission in 1992. During transformation, this institute has had to suffer the loss of half of its personnel, among them a quarter of its researchers. This has been a substantial reduction, but still below the average of all academy institutes. More importantly, it has been possible to preserve the scientific school. There have also been major cutbacks in funding, with a resulting decay of infrastructure. But no aging of researchers has occurred. In addition, cooperations with Western institutes have grown. Thus, in certain respects the institute has been somewhat successful, and in other respects it has at least managed not to fail as much as other institutes.

Other cases might be referred to which roughly corroborate the view that the scientific quality of an institute increased its chances of being relatively successful during transformation. For instance, two of the Polish cases, namely the institute of physics, and the biological institute, suggest this, as do two Slovakian cases, the physics and the pharmacology institute. All of these cases show that a high international standing and corresponding networks bring about comparatively good chances of success in the competition for project funds, especially from Western sources. This extra money in combination with collaborative international research projects has opened up opportunities for coping at least partially successfully with the trouble an institute has had to face.

The countries just mentioned – Slovakia, Poland – and the Czech Republic are also those where all academy institutes were subject to an overall evaluation which resulted in a categorization of each institute according to its scientific quality. An important consequence of this evaluation was that those institutes which were rated best have been favored in the allocation of institutional funds. Thus, this formal procedure emphasized the significance of scientific quality as a determinant of relative success or failure.[5] For example, the Polish institute of social sciences did not get high marks when it was evaluated, has since had big trouble, and has only been able to survive by teaching. Its future prospects are still uncertain. The Slovakian institute of ecobiology, whose performance rating was average, was relatively less successful than the other two Slovakian institutes, which had a better rating.

All the cases mentioned so far fit by and large with the general impression that scientific quality has been an important determinant of an institute's fate. Now we turn to those cases which demand certain modifications to this simple thesis. To begin with, there are institutes of good or even very good scientific standing which nevertheless have failed in relative terms. For example, the Hungarian chemical institute undertook research that received international attention, but has achieved nothing more than survival. Its personnel has constantly declined, its infrastructure has deteriorated, and its researchers have been forced into – what they themselves call – "brain-waste". Because of their substantial pay cuts, they have had to accept all kinds of scientifically uninteresting contract work, mostly for foreign firms, in order to earn a living. What has been especially perplexing, from the point of view of many researchers, is the fact that the institute, despite having been given a very positive evaluation, has received less institutional funds from the academy than other institutes that had been given poor ratings. As one head of department within the institute expressed it: "Our lofty position in the rank order of performance is inversely proportional to our position in financial support."

5 See also, for East Germany, Wolf's (1996, 289-290) observation that a high level of scientific quality was a *conditio sine qua non* for an institute's relative success in the West German Science Council's evaluation of the academy.

Thus, on the one hand, above-average scientific performance has not always increased an institute's chances of success. This implies, on the other hand, that mediocre institutes have not always fared less well than better institutes. For example, the Belarussian biological institute has suffered as much as, but not more than, the other two Belarussian institutes, although the scientific quality of their research seems to have been higher. Similarly, in Russia, the institute of biology, which received a high rating for its scientific quality, and the institute of radio engineering, which received a poorer rating, have not differed much in their fate.

Often, this lack of importance of scientific quality for an institute's fate is a direct consequence of the mode of institutional funding. In many countries, the allocation of institutional funds to an institute of the academy is still, as it was before transformation, tied to the number of established posts the institute has for its personnel. The more posts an institute has, the more funds it gets. This is the reason why many institute directors have hesitated to dismiss personnel, even if there has not been enough work to do, and have preferred instead to rely on other measures, such as reductions in salaries, tolerance of their personnel having second and third jobs, reductions in working hours, and imposed longer holidays. Since institutional funds are the main source of financial resources in almost all cases, and have frequently become even more important as a consequence of the difficulties experienced in acquiring project funds or research contracts, the weight of criteria of scientific quality has often been low.

Another interesting comparison is between two of the Latvian institutes. The institute of organic synthesis which conducted very high-quality research has been quite successful, but not more than the Latvian institute of physics, which had problems maintaining quality in certain segments of its research activities before transformation. The latter case is an example of an institute which quickly abandoned these unsatisfactory research activities when financial pressures intensified, thus demonstrating that active coping may neutralize the otherwise detrimental effects of insufficient scientific quality on an institute's fate.

Even more radical in this respect were the coping efforts of the Czech institute of philosophy. Before transformation it had conducted rather low-quality research. What is more, research had been ideologically impregnated by socialist doctrines. This augured badly for future survival. Yet the institute managed to quickly replace almost all of its researchers by others who were not only above suspicion politically, but often had affinities to the former political opposition, which by then had taken over power. This went along with a complete change in the institute's research topics. This radical and visible renewal of the institute has secured its survival although the quality of its research is still not above average.

There is also one case which exemplifies that an institute conducting work of rather mediocre scientific quality could succeed remarkably well during transformation, even without much of a coping effort. This is the institute of

Lithuanian language. It has experienced a moderate growth in personnel, has been able to continue with its former research topics, and has even acquired a new building together with other infrastructural improvements. The primary explanation for this success, as well as in the case of the Czech institute of philosophy, is the great importance of the institute's research for the new government's efforts to restore the former cultural identity of the country. Both institutes have been able to convince political actors of the usefulness of their work for these important political aims. This shows that even low scientific quality can be quite irrelevant to an institute's relative success or failure if the extra-scientific usefulness of its research is prominent - an observation which directs attention to the second of the supposed general determinants, the type of research an institute conducts.

The most striking contradiction to the suggested correlation between scientific quality and relative success is certainly the case of the Bulgarian institute of molecular biology. Here, scientific excellence literally has worked as a highly destructive factor against the institute.[6] The reason is that this institute has been the victim of a massive brain-drain of its best researchers to Western institutes. Thus, precisely because the institute was able to produce outstanding research results, to recruit and train very good young researchers and, in this way, achieve a high international profile, it was thrown by the transformation into the international market for researchers, where it could not compete because of the continuing and increasing lack of finances, and where it, consequently, has lost a large part of its most important productive force. Suddenly, attractive exit options opened up which many researchers could not resist.

To summarize, the case studies show that there is no simple correlation between the scientific quality of an academy institute and its relative success. Firstly, if outstanding scientific quality leads to brain drain of researchers, the correlation is inverted. Secondly, under several conditions there is no strong correlation at all between both variables: if institutional financing is guided by criteria other than scientific quality, and no significant amount of additional financial resources from project funding or research contracts can be acquired; if extra-scientific usefulness of an institute's research is high, and attention to this factor leads to a disregard for criteria of scientific quality; or if an institute is able to rapidly and visibly improve its scientific quality by active coping.

3.2 TYPE OF RESEARCH

The general supposition that an orientation towards applied, rather than basic, research will have contributed significantly to an academy institute's relative success during transformation implies that contemporary impressions from

6 The same phenomenon, but on a smaller scale, can also be seen in the Russian institute of theoretical physics.

Western countries also apply to post-socialist countries. Since the mid-seventies an increasing scarcity of resources for research has been linked with the observation that government actors of research policy as well as industrial firms have put more and more emphasis on the extra-scientific relevance of research as a decisive criterion for the allocation of financial resources to research actors. Curiosity-oriented basic research has often appeared as a luxury one could not afford any longer. This legitimation crisis of "useless" basic research, one may argue, must have been much more dramatic in post-socialist countries, where the scarcity of resources for research promotion has been invariably much stronger than in Western countries. According to this reasoning, it seems plausible that the immediate usefulness of an academy institute's research to important extra-scientific actors has been a key determinant of its chances to survive and maintain its research capacity.

In this respect, three kinds of extra-scientific usefulness may be roughly distinguished. Firstly, scientific knowledge may be an essential element of technology transfer, mainly to industry, but also to other societal sectors like the health care system or the military. Secondly, scientific knowledge may be the basis of advice given not only to political, but also to economic actors. Thirdly, scientific knowledge may serve as an important element of cultural orientation for society at large. It seems that in the current transformation period of post-socialist societies there should be ample demand for scientific knowledge in all three respects. Industry needs massive and rapid technological modernization to achieve competitiveness on the world market. Economic actors in general need all kinds of advice, such as management knowledge, to adjust to the new capitalist market economy. Political actors are in the same situation: they have to change their behavior according to the new democratic political system, in addition to being confronted with a lot of urgent and complex problems which require political decisions. Finally, all members of post-socialist societies have to learn new cultural orientations. They have the chance now both to become acquainted with Western cultural traditions and ideas, and to rediscover their own cultural past, from which they were often cut off during the socialist period. Academy institutes which produce and possess scientific knowledge tuned to these extra-scientific needs should have been able to profit from this.

Several cases, indeed, confirm this general supposition. Two cases already referred to are the Czech institute of philosophy and the institute of Lithuanian language. Both have been relatively successful because their research activities have corresponded to the societal need for new cultural orientations. Similar examples can be found in the natural sciences. In regard to the Bulgarian institute of nuclear research, for instance, the clear view is: "Partly because of the need for changes due to economic difficulties, partly because of public pressure in connection with environmental issues, in the years of transition the institute attributed much greater importance to applied studies and the policy of services. They contributed to the financial stabilisation, as well as to the improvement of its public image." Likewise, at the Russian biological institute, a "...decrease in the level of the institute's basic research ..." has been observed, and accounted for "... by reduction of governmental financing, necessity of securing the institute's survival...," the result being a "... forced reorientation towards the order not of science but market ..."

The last two examples show that the move towards more applied research is not something embarked upon enthusiastically by institutes whose former orientation was strongly focussed on basic research, even when they had not entirely neglected potential extra-scientific uses and had included them to a certain extent in the research mission (see also Darvas, in this volume). This reluctance to shift too much towards applied research originates not only from the researchers' and institutes' desire to maintain as much of their former identity as possible, which was shaped by basic research, but also from concerns that long-term dysfunctions can arise if basic research is sorely neglected. In the Russian biological institute, researchers are well aware of such problems: "At present, the institute cooperates actively enough with industrial sector R&D institutions ... which is very profitable for it. But all these are applied R&D. They bring money but no basic results. The work is oriented only to present-day requirements but not the future." "Present-day requirements" mean short-term survival, nothing more. To survive in the long term, even by means of applied research, an institute must continually renew its knowledge base through basic research which is only loosely coupled to extra-scientific uses (Schimank 1990).[7] As a consequence, some institutes have tried to find a balance between applied and basic research, which has meant that the latter has not been ignored too much, although the former has had to be intensified for reasons of survival. This can be seen, for example, in the Ukrainian chemical institute or the Polish institute of biology.

The Russian institute of economics, on the other hand, is an example where the dangers of an unbalanced drift into applied research - counselling and management training - have already materialized. The pressures of short-term survival, as well as the attractiveness to researchers of relatively high incomes earned by this kind of work, have maneuvered the institute into a dead-end situation. No young researchers have been recruited, the rate of publications has dropped, and the demand for the institute's services has already begun to fade away. The Polish institute of social sciences shows a very similar picture. It has turned mainly to teaching, giving up research almost completely. The assertion of one of its researchers that "... research will continue to be done, ... and the latest and best research findings will be used in teaching students ...", does not sound very convincing.

---

7 This is an insight which can be gained from studies of applied research institutes in Western countries - for instance, the German Fraunhofer Society.

Thus, there is a need to qualify the general supposition that applied research improves an institute's chances of being relatively successful. Institutes should not go too far in this direction because they might fall victim to scientific "burn-out". Another case in point highlights this. The Ukrainian institute of informatics had to face the situation that after the dissolution of socialism those segments of its research which had an applied character had no chance at all of survival. Before transformation, this research had been cut off from international developments by Western restrictions on transferring advanced technology to socialist countries. Now, it became apparent that the institute could not compete with Western applied research in informatics. However, during the involuntary shift to basic research it turned out that certain ideas from this segment of the institute's research had considerable potential for innovative applications. Thus, the institute's ability to cope with its situation by turning to more applied research relied on its former basic, not on its former applied, research.

In many cases, even an adequately balanced turn to a more applied research profile could not improve an institute's chances of success. The reason is simple enough. Many institutes and researchers have adopted this coping strategy with the result that the number of suppliers of applied research has grown enormously; however, the demand for this kind of research has shrunk very much during transformation. The outcome of this sudden huge gap between supply and demand has often been a dilemma which was described succinctly by one of the researchers of the Bulgarian institute of computer and communication systems: "No money for theoretical studies, no demand for practical ones." Strongly devoted to applied research before transformation, this institute later lost its established links with the computer industry when that sector, suddenly forced to compete with Western firms, faced serious trouble itself and could no longer afford to fund research. Consequently, the institute tried to go more in the direction of basic research - just as the Ukrainian institute of informatics did. But what has succeeded to some extent in that case has not worked in this one. As a consequence, the institute is now caught in a no-man's-land where neither applied nor basic research can secure its survival. This situation is typical of quite a number of academy institutes in all of these countries; what varies somewhat is only the degree of immediate danger to the institute's existence. In some cases, institutes have been able to attract Western firms. But these research contracts often result in "brain waste", which ensures survival today, but at the price of diminishing longer-term chances of survival.

In several countries, most markedly in Bulgaria and Rumania, these problematic developments have been aggravated by the academy's general policy of directing all of its member institutes rigorously into basic research. This has meant that institutes have even been hindered in their attempts to make use of those small opportunities which have opened up from time to time to earn some additional money by accepting research contracts or project funds. The funds for the promotion of basic research established in all countries do not suffice, however, to enable every academy institute to devote itself solely to basic research. If this policy is pursued to the extreme, the academy may drive some of its institutes to virtual suicide.

The strongest contradiction of the general supposition that applied research correlates with relative success are those institutes which are deeply devoted to basic research but are nevertheless - or, more precisely, exactly for this reason - quite successful. The best example of this is the already discussed Russian institute of theoretical physics. It is an institute where almost only curiosity-oriented basic research was engaged in prior to transformation. This orientation has not changed since, and the institute has been very successful. As mentioned before, this can be accounted for by the scientific excellence of the research, which secured money not only from the Russian fund for the promotion of basic research but, more importantly, from Western sources. Somewhat more ambiguous cases of the same pattern are the Latvian institute of polymer mechanics, mentioned above, and the Polish institute of physics.

To summarize this section, it is evident once again that the supposed correlation of applied research and relative success of an academy institute must be modified considerably. First, a proper balance of applied and basic research has to be maintained if this strategy is to be successful at all over the long term. Secondly, the chances to be successful through turning to more applied research depend critically on the demand for this kind of scientific knowledge; and presently this demand is certainly not big enough to guarantee most institutes' success. Thirdly, if combined with outstanding scientific quality, remaining committed to basic research may well be a more promising coping strategy for an institute to adopt.

3.3 ACADEMIC FIELD

The third general supposition, namely that academy institutes in the natural sciences have been relatively more successful than institutes in the social sciences and humanities, has partly been dealt with already. Therefore, we can be quite brief about it in our analysis here. The reasons given for this supposition are two-fold. Firstly, it is estimated that knowledge in the social sciences and the humanities has less extra-scientific usefulness than knowledge in the natural sciences. In combination with the second general supposition this leads to the conclusion that research policy makers are more likely to consider the social sciences and humanities to be a dispensable, even unaffordable luxury in these difficult times. Secondly, many fields of the social sciences and humanities have suffered another handicap in post-socialist countries. It consists in their former politically enforced affinity to socialist ideology. This ideological penetration of scientific knowledge was evident in such disciplines as history, economics, sociology and philosophy. Under these circumstances, the elimination of socialist doctrines in the research system could have amounted to a massive threat to many institutes in the social sciences and humanities.[8]

A discussion of this general supposition can start with the already established fact that the relative success of an institute is not necessarily positively correlated with the extra-scientific usefulness of its research. Consequently, even if research in the social sciences and humanities were not as useful in this respect as natural science research, by itself this would not mean hard times for the former during transformation. However, in the previous section it was shown that research in the social sciences and humanities can actually be extremely useful to particular extra-scientific actors. The Russian institute of economics and the Polish institute of social sciences have been able to earn additional money by counselling economic actors and actors of economic policy. The Czech institute of philosophy and the institute of Lithuanian language have legitimized their existence and institutional funding by their contributions to the cultural orientations of their nations. In the latter case this extra-scientific need has provided such a strong impetus that the institute will not only be maintained, but enlarged and modernized as well. Incidentally, of the four academy institutes studied that are devoted to the social sciences and humanities, none exemplifies curiosity-oriented basic research.

The problem of ideological penetration was most critical in the case of the Czech institute of philosophy. But this example shows that an institute may be able to cope effectively with such a disadvantage. As already mentioned, the institute replaced most of its researchers and changed the research topics within a short amount of time. This was possible because a "reservoir" of researchers who could be recruited for the conception and execution of the new, ideologically purified research topics existed in the environment of the institute. The case of the institute of Lithuanian language demonstrates that an institute which was ideologically barely tolerated under the socialist regime was ready to prosper as soon as this regime was abolished. Thus, even the political hegemony of socialism permitted niches for research in the social sciences and humanities which failed ideologically to toe the party line. Those institutes which had found such a niche for themselves had no problem with ideological penetration of their research after the dissolution of the socialist regime.

There are two additional aspects which have often contributed to the relative success of institutes in the social sciences and humanities. Firstly, this kind of research is relatively inexpensive, compared to the frequently highly expensive research conducted in the natural sciences. This difference may turn out to be crucial in situations of abrupt and massive reduction of available financial resources. Frequently, there has been virtually no alternative to the strong re-

8 This is at least a partial explanation for the fact that East German social sciences and humanities fared badly compared to the natural sciences in the evaluation carried out by the West German Science Council (Wolf 1996, 153-155).

duction or even dissolution of a big academy institute in the natural sciences, and simultaneously one may have conserved a smaller institute in the social sciences or humanities even though the scientific quality or extra-scientific relevance of the former was significantly higher. But the latter's budget may have been simply negligible compared to the former's. To overstate the case, closing down such an inexpensive institute may have made no difference in budgetary terms.

The second aspect which has often favored institutes in the social sciences and humanities is that they have existed in a less competitive disciplinary environment. International competition in particular is usually much less relevant than in the natural sciences. For the Czech institute of philosophy, for example, it is asserted: "The priorities of world philosophy have ... not been decisive for the orientation of research." This more local, less cosmopolitan position of institutes in the social sciences and humanities implies, above all, that the institute's scientific quality is not measured by world standards set by the most advanced institutes worldwide, but by national standards. This, in turn, may be important for an institute's standing in the perception of national research policy. An institute in the natural sciences is likely to be compared to international leaders in the respective research field – and this comparison has often been quite disadvantageous to institutes in post-socialist countries.

All in all, it turns out from the cases studied that institutes in the social sciences and humanities have not necessarily been in a much less favorable situation than institutes in the natural sciences. The extra-scientific usefulness of the formers' research may also be significant, and an ideological penetration of their research has not necessarily occurred, and, even if it has, it can be remedied. Moreover, research in the social sciences and humanities is often less expensive and not measured by worldwide quality standards.

4. Conclusion

The conclusion to be drawn from this discussion of the cases with respect to three generally supposed determinants of an institute's relative success or failure in the period of transformation is evident. There are no simple correlations of relative success with high scientific quality, a predominance of applied research, or natural science research. The causalities underlying these three apparently straightforward success factors are much more complex. Several such complicating factors which could be identified in the cases have been pointed out. Of course, in reality there will be more because the cases are surely not representative in that respect.

On the one hand, such analytical complications reduce the scope of theoretical generalizations, which also makes it less easy to predict what will happen in future to academy research in these countries. On the other hand, it is good to know that not only will strongly applied natural science research survive and eventually prosper again at some point in the future, but the social sciences and humanities and curiosity-oriented basic research will also have a chance.

At first sight, one might regret that not all scientifically mediocre or even bad research will be rigorously eliminated. Yet we know from Western experience that the evaluation of scientific quality is not always beyond doubt. More importantly, should a country give up certain important research areas altogether just because at present it has no excellent institutes in these areas, or should not a more rational research policy attempt a gradual cultivation of excellence from weakness? Both considerations remind us not to fall prey to the fallacy of elevating high scientific quality as an obligatory criterion for an institute's right to exist. Thus, even with respect to this factor, the results from the cases point to a better future than the supposed universal correlation could have produced had it been true.

References

Mayntz, Renate, Uwe Schimank, & Peter Weingart (eds.), (1995). *Transformation mittel- und osteuropäischer Wissenschaftssysteme*. Opladen: Leske + Budrich.

Schimank, Uwe (1990). Technology Policy and Technology Transfer from State-financed Research Institutes to the Economy: Some German Experiences. *Science and Public Policy* **17**, 219-228.

Schimank, Uwe (1995). Transformation of the Research Systems in Central and Eastern Europe: A Coincidence of Opportunities and Trouble. *Social Studies of Science* **25**, 633-653.

Wolf, Hans-Georg (1996). *Organisationsschicksale im deutschen Vereinigungsprozeß – Die Entwicklungswege der Institute der Akademie der Wissenschaften der DDR*. Frankfurt/M.: Campus.

ANTAGONISTIC COOPERATION IN ACADEMY-UNIVERSITY RELATIONS

EDUARD SARMIR
STEFAN ZAJAC
Slovenská Akadémia Vied
Prognostický ústav
81 105 Bratislava

1. Introduction

The changed relations between universities and academies are considered to be one of the most important effects of the transformation of the national R&D systems of the post-socialist countries. Many of the current characteristics of those relations result from the heritage of the previous organisation of their R&D systems and their principles of functioning, as well as from the effects of the socialist structure of society on scientific development in Central and Eastern Europe as a whole. How this heritage is now discussed depends, among other things, on whether the academy system of scientific organisation is understood to be an integral institutional element of the national R&D systems of Central and Eastern Europe since its very beginnings in the former USSR, or whether the academy concept of scientific organisation is understood as having been implemented by an external force as a strange and alien element which deformed the origins of the national educational and scientific systems – as with countries which became socialist republics after the Second World War, or were annexed and incorporated into the former USSR at the beginning of the '40s, viz. the Baltic republics.

In the case of those countries which were federal republics of the former USSR like Belarus, the Ukraine and Russia[1], the relationship between academies and universities during the transformation process is evolving in a less radical manner than in other post-socialist countries of Central and Eastern Europe. This gentler pace of change is evident despite the fact that important changes have been enacted within the pre-existing academies of these countries, and that new academies have also been constituted during the transitional pe-

1 The Baltic republics are generally excluded from the ensuing discussion when we speak of the ex-member republics of the USSR.

R. Mayntz et al. (eds.), East European Academies in Transition, 79–92.

riod. In the latter group of countries, both proposed and implemented changes have frequently led to conflict between representatives of the alternative research sectors; both the academies and the universities feel their interests to be threatened by changes to the system of research. However, independently of whether the change in relations has been experienced as radical or evolutionary, the reassessments of the Soviet model of scientific organisation which are taking place in all the relevant countries share many features, especially concerning mutual contacts between the universities and the academies.

In this chapter we will first describe the starting point for the change of relationship between academies and universities in Central and Eastern European countries. This starting point was rooted in the Soviet type of scientific organisation and was almost identical in each country we surveyed. In the second section we discuss the changing relationship between academies and universities in the contemporary transition period. We conclude the chapter by summarising the most important changes and continuities in the academy-university relationship.

2. The Historical Background

During the socialist period, the political authorities of all these countries aimed at concentrating research activities – particularly fundamental research – within the institutes of the Academies of Sciences. The national universities, in contrast, were regarded as institutions for the professional training of students. Although the universities have always engaged in some level of research activity, state support for the research activities of the academies was considerably greater than that offered in support of university-based research projects. The preference of the political authorities for research carried out by the academies was reflected in the fact that funding the national academies took priority over funding all other kinds of scientific organisation. This led to the creation of better conditions within the academies for scientific work to be pursued.

At the same time, universities were offered few incentives to engage their teaching staff in research activities. Heavy teaching loads weakened the link between teaching and research in a number of areas. This situation damaged the contents of what universities could teach their students, and students' knowledge acquisition over the course of their university studies was diminished. Most of the respondents in our case-studies stressed that the universities had a comparatively weak research infrastructure, that university staff in general had fewer research qualifications than their academy peers, and that the research output from the universities was low compared with that of the institutes of the academy.

To complete our understanding of the relationship between the universities and the academies in the previous regime, we must also bear in mind that all scientific activities were carried out under the omnipresent gaze of the state and its centrally-planned bureaucracies. In this context, a very important feature of the Soviet regime with its planned and directed management of the R&D systems was that research activities were very strictly compartmentalised (Gaponenko 1995). Each component of a research effort came under the remit of certain administrative departments within the various ministries – the academy was itself considered to be such a department – each operating with state-funded budgets. It was only in the case of the economic departments that research facilities and efforts were funded from their own resources. The use made of budgetary resources was linked to the activities carried out within the departments. Thus, departmental managers were in a position to shape the selection criteria for funding research projects in ways which did not necessarily reflect purely scientific criteria. Ultimately, departmentalisation caused the scientific communities' natural structures of contact with each other to erode – including those between the national academy and the universities. This hindered the realisation of joint projects and curtailed the mobility of experts from the university to the academy and vice versa. Even the common state plan did not provide a basis upon which genuine joint research activities could be embarked upon.

This does not mean that the academies and universities were completely isolated from each other, without any forms of linkage and co-operation between them. For instance, many leading university scholars were members of the academy; academy specialists participated in teaching programmes; and representatives of both kinds of institution were co-members of various management and scientific committees of the academy and of the university. In addition, the system was characterised by a great deal of informal networking, based on personal contacts, which became a significant means for researchers to sidestep the restrictions of the rigid administrative separation of the two kinds of institution. Nevertheless, in spite of these links, the relationship between the academy and the higher education institutions was one where both were trapped in a highly prescriptive administrative and financial web. Amongst other consequences, this reduced the efficiency of the research efforts within both kinds of institution.

One of the more fertile fields of co-operation between the academies and the universities was that of postgraduate studies. These were organised by both kinds of institution and they took a similar form regardless of whether they were run by the universities or by the institutes of the academies. The so-called aspirantura study marked the conclusion of a science degree, while the CSc was more-or-less equivalent to the PhD qualification offered by the OECD-member countries. A further level of postgraduate studies was called the "big doctorate", and this led to the degree of Doctor of Sciences (DrSC), which was more-or-less equivalent to the level of a full professor in the OECD-member countries. While science degree candidates had to prove their capability at engaging in scientific work, doctors of the sciences had to demonstrate that they had advanced the body of new scientific knowledge as a consequence of their research efforts. In practice, the CSc degree was only awarded to those who had passed the prescribed examinations, who had gained their university diplomas, and who had successfully defended their theses. The aspirantura thesis required that, through it, candidates could demonstrate their ability to pursue original research. The degree of DrSc could only be awarded to those who already held a CSc degree and who had prepared an additional doctoral thesis which demonstrated that they had made an exceptional further contribution to the field of science they worked in. It was required that the research results of a doctoral thesis, at the very least, should be worthy of publication in the international science literature.

In general, with the exception of Hungary, co-operation between both types of institute in offering this level of scientific training was unproblematic. In Hungary until the 1980s, however, the administration obstructed universities' efforts to establish postgraduate study programmes. In all other countries, common commissions of the academy and the universities were instituted to control training, evaluate results and confer academic degrees. Both kinds of institution provided training facilities for candidates regardless of whether those candidates were based in the universities or in the institutes of the academy.

Nevertheless, in general, the mutual relations between both institutional forms were far from good. Functionally separated as they were, the scientific potential of both the academies and the universities was undermined by the fact that the conditions for engaging in mutually beneficial co-operative (or even competitive) research were so unfavourable. Despite various informal networks within the scientific community itself, the formal relationship remained stiff and distant and without any prospect of improving.

3. The Changing Relationship Between Academies and Universities

Following the political overthrow of the old Soviet regime and the implementation of the first transformative measures which have come in its wake, the main areas of co-operation between the academies and the universities have remained the same. They still sometimes join forces in offering doctoral and postgraduate studies, teaching and supervising students, organising joint projects or conferences, and jointly preparing articles for publication. They still share some facilities for information search purposes or for gaining access to scientific apparatus and they are each members on each others scientific boards, etc. But from a systemic point of view, some of the characteristics outlined in the last section have significantly changed as a result of the re-orientation of the science policies of the post-communist countries of Central and Eastern Europe. Of particular interest here is the important conceptual shift which has occurred in re-defining how the various national R&D systems should now operate. This conceptual shift, which is embedded in the processes of transformation, alters the role of both kinds of institution – the academies as much as the universities – and implies that their mutual relationship within the national R&D system is subject to profound change. For this reason, the following sub-section considers how state policy has changed with respect to how the academies operate. This can be observed by reviewing how the scientific community now regards both the academies and the universities, and their interactions with each other. In the subsequent sub-section, we describe the new forms of co-operation between both kinds of institution.

3.1 HOW RESPONDENTS ESTIMATE THE EFFECT OF STATE POLICIES ON THE ACADEMIES AND UNIVERSITIES

From the outset of the transformation process, a redefinition of the position of the academies vis-a-vis the universities was lobbied for by the university sector in particular, and by the scientific community in general. As we pointed out in the introduction, how the re-structuring was conceived took a more radical form in those countries which were not ex-republics of the former Soviet Union; the most radical conception of all – pursued by the Baltic republics – simply wanted to get rid of any evidence of a Soviet-like heritage.

Representatives from the academy institutes in all countries share the opinion that the universities need greater state support and deeper co-operative links with the academies. Obviously, this basic idea was – and is – conceived differently across different countries. The Baltic republics, Czechoslovakia and Poland (in the earliest stages), displayed a strong desire to abolish the academies and to transfer the academies' institutes into the university sector. In Lithuania and Latvia, the academies were, in fact, transformed into learned societies, and their research units became autonomous institutes (see also Simeonova, this volume). In these countries, it is proposed that the academies' institutes should eventually be completely integrated into the university structure although, to date, there are no specific programmes for implementing the transition to this new form of organisation. Opinions on this issue are very heterogeneous. For instance, a respondent from the Latvian institute of organic synthesis stressed that the likelihood that this institute will come under the umbrella of the university sector sometime over the next few years is slim indeed.

The change in the academies' role and status has evolved differently in the other post-socialist countries. Here, the academies remain an intrinsic part of the national scientific landscape – albeit with reduced competencies. The academies' institutes now are empowered to supervise their own research units, and they enjoy greater autonomy in controlling their own research activities. Of course, the position of the academies is different in each country. However, a general tendency observable in all countries is that the higher education sector has enjoyed more advantages (in terms of improvements in material conditions and in personnel) than has the academy sector, as a result of the change in policy orientation.

Academy staff numbers have declined more than university numbers have. Indeed, in Estonia, the Czech Republic, Bulgaria and Romania university staff numbers have increased. Despite an overall severe reduction (in real terms) of monies allocated to R&D activities in all post-communist countries, most respondents considered that the universities are currently in better financial health than are the academies. Generally speaking, and with the exception of Hungary, university funding has increased over the last few years. In Hungary, the government made unprecedented budgetary cuts in the middle of 1995 which compelled the universities to dismiss thousands of their staff. This made it impossible for academy researchers to maintain their involvement in university teaching programmes there. In the Slovak Republic, funding was also cut in 1995 whereas in previous years it had increased (nominally) on a regular basis. In Estonia, interviewees claimed that the universities were not privileged by the funding authorities.

In Poland, the higher education institutions compare their share of the available R&D budget unfavourably with the share the institutes of the Polish Academy of Sciences (PAS) receive. Yet according to respondents in the PAS institutes, they have not been favoured in the distribution of funds. Their share of the pool of funds, which is allocated by the Polish State Committee for Science Research, has decreased each year. Most of the members of the State Committee for Science Research are representatives of higher education institutions and it is they who are responsible for the distribution of the financial resources. It seems likely, however, that the problem is not one of fierce competition for grants between the two sectors nor one of systematic unfairness in the distribution of grants. Rather, the core difficulty is that the available funds are meagre, and both sectors suffer a budgetary squeeze as a consequence.

Nevertheless, the general opinion expressed within the framework of the survey is that – with the exception of Hungary after 1995 – the current working environment in the universities is more stable than it is in the academies. It is interesting to note that this situation has not led to an exodus of researchers from the academies' institutes into the universities. Mobility in this direction remains low, despite the more stable and advantageous situation the universities currently enjoy relative to the academies. (In addition, scientists' mobility from the universities into the academies also remains low.) However, in the case of Hungary, the political preference given to the universities in the early '90s coupled with the uncertain future of the academy's network during the same period attracted several leading scientists to leave their institutes for university positions. Romanian respondents also noted that higher education institutions have become more attractive to scientists as there is still a high demand for teaching staff and the universities offer greater workplace stability, together with the chance to engage in doctoral studies both at home and abroad. However, on the whole and across most of the post-socialist countries, the mutual mobility of qualified personnel between academies and universities remains weak.

Despite this situation, the academies' researchers recognise the value of interactions between the academy and the higher education institutions; they even recognise that the national R&D systems will gradually become more integrated. However, it is only in the Baltic republics that this process of integration has already taken place. In the other countries, the strategic integration of academy and university structures is regarded as an important – but non-urgent – aspect of the successful transformation of their R&D systems.

It is more common for the question of the integration of the academies and the universities to be conceived as a process of step-by-step concrete efforts to deepen co-operation between them, e.g., to conduct joint projects, to establish laboratories for common use and to share each others facilities where appropriate. For instance, in Hungary, the Higher Education Act and the Act on the Hungarian Academy of Sciences (HAS) have made it possible for common associations to be formed. The HAS institute of chemistry and the Faculty of Chemical Engineering of the Technical University of Budapest founded just such an association, involving research staff from the institute's associated department and university research teams. In October 1994, the joint Polymer-Physics Laboratory – which is equally equipped to engage in research and educational activities – was constituted as the first result of this association. Common laboratories also exist in other countries. Russian, Ukrainian, Lithuanian and Czech respondents mention similar developments. Depending on the specific national context, we also find instances where an academy's institute operates within the framework of a university structure. Such an example is provided by the Ukraine although, in this case, this particular academy-university interaction was implemented during the communist period, and is not an outcome of recent developments in the Ukraine.

A notable development is that the academies' institutes have begun to offer their own teaching programmes. This was the case in Lithuania, where the institute of biochemistry founded a chair of biochemistry within the University of Kaunas. Even in Russia, the academy's institutes have begun holding chairs of their own at higher education institutions. In Poland, a School for Exact Science operates inside the framework of the PAS. This school was established by five PAS institutes together with one ministerial institute as a non-public-sector enterprise offering an interdisciplinary educational programme. In October 1992, an Interdisciplinary Postgraduate School for Social and Human Sciences was also established within the PAS framework. Another example of a PAS institute's involvement in teaching activities is the creation of a PAS-owned college (also a non-public-sector school), which confers professional degrees such as the MBA.

Despite these sorts of local developments, the great majority of survey respondents in all countries reject the idea that the academies' institutes should be integrated into the universities – a restructuring which would make them subordinate to university authorities. According to most respondents, any integration of the scientific activities of academies and higher education institutions must be conceived in terms of establishing complementary co-operative links rather than in terms of abolishing the academy network or subordinating it to the overall authority of the universities. The Romanian respondents expressed this argument most precisely. According to them, given the scientific prestige of the academy network, the dismantling of the academy's research institutes or their transfer into the university structure would mean:

a) the Academy of Sciences would simply disappear if the network was dismantled. In short, academy science is seen as part and parcel of Romanian intellectual culture;
b) a serious migration of research staff who lack pedagogical talents;
c) the risk that scientific research would become marginalised and disadvantaged, in favour of the teaching activities which the universities have traditionally prioritised and rewarded in terms of funding.

Belarussian interviewees also emphasised that some potential advantages of combining the academies' institutes with the universities along the lines of the Western model of scientific organisation – for example, reducing the administrative bureaucracy; involving students in research activities; and augmenting the training of both students and their teachers – are neutralised by the risk that excellent academy-located research teams may be destroyed in the process.

The idea of a simple administrative transfer of academy institutes into higher education structures does not enjoy the unconditional support of university personnel either. Slovak respondents pointed out that universities are unable to assimilate the research institutes of the academy, even if it were guaranteed that all the institutes' finances would be transferred with them. One cannot disregard the fact that the universities generally have a large – and hitherto underused – bank of research potential amongst the ranks of their own teaching staff. Due to their higher scientific prestige and achievements, academy researchers may supplant many of the universities' scholars. And all this would happen in a situation where the funding of R&D activities has been radically reduced and the future prospects for financing scientific research remain uncertain.

Moreover, respondents from the institute of molecular biology of the Bulgarian Academy of Sciences (BAS) assume that university lecturers fear the competition they would face from BAS scientists who might qualify as candidates for professorial vacancies within an integrated system. An interviewee in the Latvian institute of physics also claimed, in connection with this issue, that the universities profoundly distrust academy researchers; so great is the threat that the latter pose. Thus, even if they wished to enter the university system, academy researchers would find it difficult to by-pass the barriers erected against them by university staff.

We can conclude that in the academic R&D activities of the post-socialist countries a new situation is emerging which is having the effect of redefining the roles of the academies and of the universities and, in the process, is generating new forms of mutual relationship between them. Under the old system, the academies dominated fundamental research activities and enjoyed a preferential status in the eyes of the political authorities, while the higher education institutes were consigned to playing second fiddle. Accompanying the process of economic, political and social transformation in all the post-communist countries, we find a conceptual shift taking place which argues for the equality – if not the primacy – of university scientific activity. This shift reflects the increased support the political authorities now offer the universities (including support for their research activities). With the exception of Hungary since 1995, the result is that the universities have become the more stable work environment of the two kinds of R&D institution. The Western model of university science is often invoked as an argument in favour of the radical restructuring of scientific activities; in some cases, the argument even stretches to include the proposal that the academy network should be dismantled.

The scientific community, whether located in the universities or the academies' institutes, agrees that it is necessary to raise the quality of university research. Towards this end, scientists agree that closer co-operation, even to the point of integrating the research activities of both sectors, is highly desirable. However, the great majority does not conceive integration in terms of dismantling the academies. Thus, while almost all respondents seek closer mutual relations, the preference is that closer co-operation should be achieved gradually, as the natural and organic outcome of a series of concrete co-operative agreements between the two. While there is evidence of several such concrete steps being taken – an issue we will explore in the next sub-section – not all of them have significantly improved the relationship between the academies and the universities.

3.2 NEW FORMS OF CO-OPERATION BETWEEN ACADEMIES AND UNIVERSITIES

3.2.1 *Co-operation in the Field of Doctoral and Postgraduate Studies*

During the transformation of the R&D and higher education systems in Central and Eastern Europe the academies have continued to engage in postgraduate study programmes. However, in almost all of the post-communist countries, the uncertainties of the transition period have meant that scientific careers are currently less attractive than they previously were. In consequence, respondents note that there is a decline of interest in such studies. In addition, how these studies are conceived differs across the countries we surveyed.

In Russia, the Ukraine and Belarus, the previous aspirantura system remains unchanged. This means that the academies' institutes as well as the universities are entitled to provide such a programme of postgraduate studies. In Poland and Lithuania, which model their system along Western lines, doctoral training can be offered by both the academies' institutes and the universities. In Estonia, depending on the availability of qualified supervisors and of suitable facilities, both types of research institution offer both doctoral and masters training programmes. In Hungary, Latvia and Romania, doctoral training is offered exclusively by the universities although the academies' institutes can provide additional back-up (for example, access to laboratories and information sources), and academy researchers may act as co-supervisors. In the Czech and Slovak Republics, the academies have run the aspirantura study programme since 1989. However, new doctoral study programmes, introduced since 1990, fall to the universities to provide. This means that there are two systems of postgraduate study in these countries which run in parallel. The Slovak Republic's Act on Higher Education aims to end this situation. The Act proposes that all doctoral studies should properly belong to the higher education sector but that postgraduate training should be supervised by a scientific board whose members are drawn from both the universities and the academy.

3.2.2 *Co-operation in the Higher Education Training Process*

The (partial) involvement of academy researchers in the universities' educational activities is a second area where the two sectors have traditionally co-operated with each other. The majority of academy scientists regard supplementary lecturing as part and parcel of their scientific work, and are happy for this connection with the university system to be maintained or strengthened. Our survey indicates that, in practically all post-communist countries, the participation of academy scholars in university teaching has actually increased during the transformation period. This seems especially marked in the Czech Republic, Estonia, Hungary, Poland and Romania. In other countries, participation levels have not increased, nor have they decreased over the same period. However, in a third group of countries – Bulgaria, Latvia and the Slovak Republic – participation levels have fallen. A Slovakian respondent from the SAS's institute of ecobiology stated that the universities are apprehensive of SAS involvement and actively restrict the participation of academy researchers in their educational activities. Similarly, a Bulgarian respondent from BAS's institute of nuclear research noted that teaching work has became largely inaccessible to academy researchers there.

3.2.3 *Co-Operation Through Joint Projects or Joint Publications*

Levels of co-operation in this field also differ across different countries. For the most part, the amount of joint work remains unchanged from that which pertained under the old regime. In some cases, the level has improved. The Czech Republic provides just such an instance, where the introduction of a grant funding system has had the effect of increasing the number of joint projects and publications. However, in the Slovak Republic, common grant projects have been beset by administrative difficulties and are not hugely popular as a result. In Hungary – where, historically, the academy has always maintained good contact with the universities – the amount of joint endeavours has also increased during the transition period. In contrast, Lithuanian and Russian respondents note that there are practically no joint projects running in these countries and, even where they do, the numbers remain paltry. Respondents from the Ukraine stress that the number of joint projects has actually decreased in recent years.

In Belarus, joint projects are conducted under the aegis of the national Science & Technology Programme. However, as the Belarussian respondents point out, university researchers only participate in (nominally) integrated research programmes in a formal sense. Such enterprises are directed by the academy's institutes and the actual research programme is typically divided into discrete projects, each of which is separately done by the participants in each sector. As with the Slovak Republic, it would appear that the funding mechanism continues to endorse the compartmentalisation of research activity into discrete spheres of influence. However, Belarussian respondents have also noted that the increased opportunities to participate in international projects is having the effect of stimulating local academy and university scientists into setting up temporary joint research teams in their pursuit of international funding. This new trend is also apparent in other post-communist countries in our survey.

Polish respondents have pointed out another interesting new feature about joint endeavours which augments the comments of the Belarussians about the effect new variables are having on the status quo. In Poland, co-operative links between the academy and the universities are qualitatively different since 1991. Previously, entire research teams from the respective sectors engaged in joint work. Since 1991, however, the focus has shifted to individuals. That is, an academy scientist may now choose to co-operate with another scientist (who may be based at a university) for the purposes of doing joint work – but the choice of partner will be based on criteria of scientific qualifications, shared professional interests and informal knowledge of each other. This trend towards the individualisation of co-operation in research work can also be observed in other post-communist countries and signals the emergence of a qualitatively new feature in the organisation of research activity.

3.2.4 *Mutual Use of Research Facilities*

The mutual use of research facilities, especially with respect to access to the technical research infrastructure, has not significantly changed. The academies' institutes – which are usually better equipped than the universities – offer university researchers access to their facilities. However, Belarussian respondents point out that whereas the academy's institutes used to offer scientific and methodological assistance to university teachers over the probation period of their employment, this kind of co-operation is on the decline. The same can be said for Estonia and Latvia where university teachers use the academies' institutes to gain research skills. In Hungary, as in some other countries, the academies' institutes offer graduate students access to their laboratories so that students may complete their projects. As we outlined in Section 1, this kind of co-operation marks a continuation of the Soviet style of organising the R&D system. Although we found some departures from this heritage in some of the countries we looked at, the situation in most cases remains unchanged in this respect.

4. General Conclusions

By and large, the relationship between the academies and the universities has not dramatically changed in the transition currently being experienced by the post-communist countries. In most countries, the academy network has survived intact, despite the strong criticism levelled at it by the universities. Academy researchers continue to partake in the teaching activities of the universities and most academies continue to confer the doctoral degrees of CSc and DrSc. In most cases, academy institutes co-operate with universities in doctoral training programmes. In several countries, joint research units continue to function and academy and university researchers can, and do, collaborate on joint projects.

Nevertheless, despite these continuities with the old system, the picture is not as homogenous as it would first appear. Behind the apparent stability of the relations between academies and universities, we find new departures in several countries in the way co-operative links are formed. These departures from past practice suggest that a substantial shift is underway. With the exceptions of the Czech Republic, Latvia and (to some extent) Hungary, the relationship between the two sectors is becoming weaker in quantitative terms. At the same time, however, it is also becoming more spontaneous and informal than was possible under communist rule. This spontaneity does not necessarily or always contradict the fact that there has also been an increase in some countries in the number of formal co-operative agreements being signed between the two sectors. Nor does this spontaneity imply that the quality of work produced through joint projects or shared activities has deteriorated.

Many respondents stress the decisive importance of personal contacts – contacts which were forged under the old regime but which have been maintained and enhanced in the transformation period – in enabling scientists from both sectors to engage in mutually beneficial co-operative endeavours. It is the existence of this informal network which offers scientists the means to side-step some of the obstacles which tenaciously remain in place as a result of the historic compartmentalisation of the funding and organisation of research activities. In addition, the entry of new opportunities – most notably, the opening-up of opportunities to engage in international research projects – contributes to the reshaping of the scientific landscape. The fact that scientists can more easily participate in international projects is having the effect of enhancing the local formation of research teams composed of scholars from different R&D sectors, most especially from the academies and the higher education institutions.

The reasons for these new developments can be traced to the political overthrow of the old regime. By the beginning of the 1990s, the principle of legally-guaranteed autonomy had permeated, inter alia, to the realm of scientific research in all the Central and Eastern European countries. Both the academies and the universities gained some measure of autonomy to make decisions on their own behalf at the same moment in history when the ensuing economic crisis severely affected their funding. In this context, both kinds of institution have been compelled to seek out for themselves the most rational ways of deploying the financial resources they do have. Despite the limitations that budgetary cuts impose, the R&D institutions have been forced to become more creative about searching out research opportunities and more motivated to ensure that consistently good work is produced. As a result of the latter pressure, many less productive researchers and technicians have left the academies' institutes. As a result of the aforementioned principle of autonomy, in most of the countries we surveyed researchers themselves can choose which projects they wish to pursue, and with whom they wish to forge research partnerships.

Nevertheless, the chronic lack of resources has had a negative impact on certain forms of co-operation between academies and universities. Both sectors tend to want to preserve the facilities afforded by their own workplaces for use by their own staff, regardless of their professional capabilities to make best use of their technical infrastructures. Some survey respondents noted the guarded stance some of the universities take towards teachers who also possess advanced research skills. What is most alarming is the drop in numbers of those wishing to undertake doctoral work, as a result of the unstable and unfavourable research climate which has accompanied the transformation period (see also Nesvetailov, this volume). The decrease in doctoral candidates has a deleterious effect on this very important field of co-operation between academies and universities and, of course, on the prospects for scientific activity in the post-communist countries in general.

The majority of the survey respondents stressed that the academies and universities are not substantially in competition with each other. Each enjoy their own sphere of influence, be it in education, research or a mix of both. Nevertheless, rivalry persists between them in pursuit of scarce research funding sources, especially in situations where the grant mechanism of distributing funds operates. In all the surveyed countries, we found no increase in the mobility of personnel between academies and universities. Although the academies' institutes cannot provide their staff with a stable work environment nor offer high salaries (and on these grounds a high turn-over rate might be expected) the universities are not, in general, open to the idea of being invaded by ex-academy researcher-lecturers.

More positively, we also find evidence of some deliberate strategies to enhance academy-university links, such as the setting-up of common laboratories. In the Ukraine and the Czech Republic, there have been attempts to create a joint research infrastructure by establishing unified library funds. In Estonia, an Inter-University Information Centre has been opened. In almost all countries, students and university scholars have access to the academies' laboratories for the purposes of postgraduate and practical research training.

The most positive development of all, perhaps, is that the majority of the academies' scientists are in favour of the gradual integration of the academies and universities – though this basic endorsement is conceived of differently in each country. Most see integration as a step – or a series of steps – which can only be taken well into the future, and that attempts to unduly quicken the pace of change will doom the whole enterprise to failure. In turn, the political authorities of almost all the post-communist countries seem well-disposed to support gradual integration through their S&T policies.

The present state of relations between the academies and universities in the post-communist countries of Central and Eastern Europe should be regarded as transitory. The current situation reflects the desire to deepen the connection between teaching and research for the benefit of both activities, as well as the influence of the very complex economic, social and political turmoil which the ex-communist countries are currently enduring. We may hope that when the economic situation improves sufficiently, then many of the obstacles which currently prevent deeper co-operative links being forged will be removed. In that future scenario, the academies will become – if not an integral part of the university system – then its natural scientific background.

References

Gaponenko, Nadezhda (1995). Transformation of the Research System in a Transitional Society: The Case of Russia. In: Katalin Balázs, Wendy Faulkner, & Uwe Schimank (eds.), *The Research System in Post-Communist Central and Eastern Europe* (pp. 685-704). *Social Studies of Science* **25**, EASST Special Issue.

COMPROMISED FUTURES: THE CONSEQUENCES OF AN AGING RESEARCH STAFF

GENNADY A. NESVETAILOV

CIS
Academy of Sciences of Belarus
220 600 Minsk

1. Introduction

One of the grave problems created by the transformation process is the imbalance in the demographic structure of the research system. An imbalanced age structure threatens the continuity of research productivity and development. Some figures are truly dramatic. In the Ukrainian institute of physics the percentage of researchers younger than 35 made up 23. 2% of the total number of researchers in 1990 but as few as 10% in 1994. In the Russian institute of economics there are about 230 researchers, half of whom are older than 50 and only 16% are younger than 40. In the Hungarian institute of solid state research researchers older than 40 constitute 60% of staff while only 10% are younger than 30.

The imbalance in the age structure has resulted from both external and internal factors. The aging of research personnel, in turn, has brought about a number of consequences for the organisation and productivity of scientific activity. A comparative analysis of the case studies reveals that the aging of researchers has both general and specific features. In this chapter, I will first briefly discuss research results on the significance of age and the age structure of research personnel, then look at the data from the case studies and finally report on some strategies which institutes have pursued to cope with the skewed age structure.

2. Individual Aging

The effect of a researcher's aging on his or her professional activity has been the subject of scientific debate (Simonton 1988; Zuckerman & Merton 1972). The American psychologist H. C. Lehmann who spent 25 years working on his book

R. Mayntz et al. (eds.), East European Academies in Transition, 93–106.

Age and Achievement (Lehmann 1953) found that the productivity peak and the highest quality of activity occurred between the ages of 30 and 39.

The problem of age and creative activity encompasses at least three sub-problems: age and great scientific achievements; age and the productivity of the average researcher; and age and the development of new scientific fields. Regarding the first sub-problem, a statistical analysis of Nobel laureates is one of the most popular methods of ascertaining the validity and the effects of this issue.

For example, for an array of 414 Nobel prize winners in the period 1901-1992 the age distribution is unimodal in all of the three sciences of physics, chemistry and medicine (Levin, Levin & Levin 1993). The peak creative age is 31-35 in physics, 36-40 in chemistry and 31-35 in medicine. In all three sciences, 85% of the winners reached their creative peak at an age younger than 45. Thus, it is recognised that the greatest potential for producing important research results comes from young scientists. These results were, in fact, taken as normative standards by academic institutions in the countries of Central and Eastern Europe for guiding their manpower policies. Under the new conditions of transformation, those academic institutes which remain centres of excellence may still operate by this rule of thumb: 68% of the scientists in teams doing basic research in the natural sciences should be younger than 40.

The question of aging and creative activity becomes an issue that is more or less important to address depending on whether outstanding or 'normal' research results are at stake. H. Lehmann claimed that the creative peak appeared sharper for the most outstanding science achievements, but that for less important research outputs (articles, monographs, patents, etc.) it is flatter (Lehmann 1953). Sociologists have carried out studies to verify this hypothesis (Pelz & Andrews 1996). For example, the results of the international ICSOPRU project confirmed the appearance of two peaks of publishing activity at 45-49 and at 55 years for researchers at academic institutions in the natural sciences. This showed that an average researcher's productivity also depends on his or her age, even though this dependence is not so pronounced as it is for outstanding research achievements.

As to the third problem – the link between age and the development of new research fields – it is known that two factors are important for successful research: the amount of acquired knowledge and tradition (i.e., 'wisdom') and the ability to consider and deploy fresh approaches (i.e., 'flexibility') (UNESCO 1979). As time passes, more wisdom is gained and more flexibility lost, and the researcher becomes less receptive to new ideas. Conversely, the younger the researcher, the more he or she is inclined to assimilate the fundamental paradigms of science. According to Thomas Kuhn, "Almost always the men who achieved the fundamental inventions of a new paradigm have been either very young or very new to the field whose paradigm they changed… , this generalisation about the role of youth in fundamental scientific research is so common as to be a cliché" (Kuhn 1970).

It seems to be an inherent property of the nature of scientific ideas that the new generation are an important vehicle for scientific progress. The conflict between new and old views, routine approaches and new paradigms, is always influenced, on the one hand, by particularly outstanding scientists – the authors of ideas – and, on the other, by young researchers who easily adopt new concepts. There have been some empirical studies of the influence of aging on the ability to perceive or generate new scientific ideas (Messeri 1988; Stewart 1986). The readiness of the young to quickly adopt new research trends has been demonstrated convincingly in the instance of neural networks research (Rappa 1993). Thus, the aging of research personnel seems to be closely related to a cognitive aspect of research activity, i.e., the readiness to adopt new knowledge. The dynamism of scientific progress hinges on the continued advent of new generations of researchers.

3. Changes of Personnel Age Structure in Institutes

Two types of the collective aging of research personnel can be distinguished. The first type is based on the life-span of a research team, and the second is determined by the (physiological) age structure of the personnel. In the former case, the collective age is calculated as the average number of years each member has worked in that research team; in the latter case, it is defined as the average age of research staff within, for example, an institute or across the entire national scientific community.

There is a particular connection between both types of collective aging. The older researchers grow, the higher is the probability that they belong to research groups that have been in existence for a long time. The age of the research group affects its productivity as was demonstrated by the international ICSOPRU project (Stankiewicz 1979). Since the case studies project did not aim to collect or analyse the age data on research groups, only the second type of collective aging will be considered here.

The age dynamics of a research team have much in common with the individual aging of researchers. Just as is the case with individuals, every team (group, laboratory, institute) experiences four developmental phases over time. However, collective aging can be controlled which cannot be said for individuals. In particular, control of the average collective age is of great importance. In research institutions, it is necessary to take some measures to stabilise the age structure – especially under conditions of restricted financing and a reduced inflow of young researchers.

It seems almost paradoxical, though, that intense personnel turnover may be beneficial for pursuing a personnel rejuvenation policy. Indeed, against the general background of the distorted personnel age structure, some institutes were found to having rejuvenated their personnel. In particular, in a study of 49 S&T organisations in Russia, 46.9% of discharged researchers, but 69% of newly-hired researchers were younger than 40 (see Analytical Centre for Science and Industry Policy 1996, 80-83). The percentage of people younger than 25 constituted 4.5% of the discharged researchers and 29. 3% of the newly-hired scientists. Researchers older than 60 amounted to 2.7% of newly-hired staff and 5.3% of discharged scientists. Thus, under transformation, the turnover of research personnel can, in principle, favour its rejuvenation. Unfortunately, however, the experience of the institutes studied testifies to a collective aging of scientific staff rather than to its renewal.

4. Aging and its Consequences for Research

A reduction in the number of young scientists and a consequent distortion of the age structure has affected the research systems in the transforming countries in ways which are considered to hold poor prospects for the future. The concern about this development is deep and widespread. Thus, for example, in the Lithuanian institute of semi-conductor-physics, 85% worried about the decreased influx of young researchers and 89% about the aging of the research personnel in general. Respondents were unanimous in their opinion that it is not the reduction of personnel per se which is important but the cessation of an inflow of young scientists (Polish institutes of physics and biology). The low social prestige of science discourages talented university graduates from embarking on an academic career. Whereas prior to the transformation period academy institutes were able to choose from a large number of applicants, currently some have not been able to hire a single young researcher.

University graduates and postgraduates have always been the principal personnel reserve for research teams. If they are no longer available, this must have the consequence of reducing the number of researchers. At the Bulgarian institute of molecular biology vacancies remain unfilled, despite a high unemployment level among highly-educated young people. Respondents in the Czech institute of chemistry consider that the shortage of young scientists, coupled with the need to maintain their success rate in winning grants, will probably result in the institute having to employ researchers aged 60 years and above. This was at least foreseen for 1995 and 1996.

Everywhere weak links between the academy and the universities remain another important reason for the reduced inflow of young people into science (see also Sarmir/Zajac, this volume). This negative trend seems to be most distinct in the former Soviet republics. Thus in Latvia a respondent from the institute of physics noted that "universities are surrounded by an almost insurmountable fence." However, even institutes which have developed close contacts with universities face the problem of replenishment. For example, at the Hungarian institute of solid state research, about 28% of the researchers teach at universities. Special funds are allocated annually to the institute by the Presidium of the Academy for the purpose of hiring young researchers. Nevertheless, there are few young employees at the institute. Researchers older than 40 constitute 60% of total personnel whereas those younger than 30 amount to as little as 10% of the total.

This situation is aggravated by economic conditions. The drain of young researchers abroad is observed at the undergraduate level for it is easier to live 'there' on a scholarship than 'here' on a salary. Thus, the potential reserves of integration between academy institutes and the universities should not be overestimated. Remuneration of researchers and, thus, the material prestige of scientists' professional careers remains a crucial issue.

A decrease in the number of postgraduates and of defended dissertations is a typical consequence of the reduction in the number of young researchers. For example, at the Polish institute of physics the number of defended dissertations decreased from nine in 1991 to one in 1994. At the Belarussian institute of physics there were 31 postgraduates in 1990 and as few as nine in 1995. In the same period, the number of defended dissertations fell sharply: from 14 in 1990 to two in 1995. As a result, some respondents look pessimistically at the training of highly-qualified research personnel: "It's a complete failure. It can't be repaired. Everybody understands that something must be done, but there are no funds."

For the insignificant number of young people who still come to academy institutes the 'came, looked around, left' pattern has become a routine. Typically, for them, a research institute is a transition site, where they can upgrade their professional qualifications in order to engage subsequently in non-scientific fields such as business and commerce. This was evident in the complaints made by interviewees from the Ukrainian institute of physics: "Today there is the following situation at our institute. Young university graduates are hired. They are clever, bright, motivated and start from postgraduate work. They have an action plan and they try to follow it strictly. During the postgraduate work they seek possibilities for practical activity or foreign contracts. At the end of postgraduate work, they present their dissertation and leave the institute for business, commerce or trade". In 1994, 25 graduates were hired and 23 young researchers left this institute. Thus, even the institutes that are able to attract young people cannot retain them and a noticeable level of collective aging is observed there. In particular, at the Ukrainian institute of chemistry the percentage of age groups younger than 30 fell from 27.5% in 1990 to 19.3% in 1994. The percentage of researchers at a pensionable age simultaneously almost doubled.

All this results in collective aging which, in turn, decreases the probability of being at the cutting edge of new trends in basic research (Nesvetailov 1986).

The director of the Hungarian institute of chemistry holds that only much younger research personnel can help the Institute develop advanced projects which are of international importance. Like scientists at many other research institutions, leading researchers at the Czech institute of physics are aged about 50. This has produced pessimistic predictions of the possible development of new areas of research: "They earned recognition and success in their fields and one cannot expect them to change their specialisations and to focus on new topics as would be required by the development of research in physics". Interviewees in the Russian institute of radio engineering and electronics point out that a dearth of young people has a negative effect on the work of researchers of medium and higher ages.

In most institutes, the brunt of responsibility for developing research trends is still largely borne by the older generation of scholars. They give stability to the system in conditions of economic crisis. But most of them will soon be pensioned off and their replacement is of great importance to most of the institutes studied. The main problem is to ensure a smooth exchange between the generations, especially in view of the need to direct research towards new and promising themes. But, at many institutes, the succession of generations in science will be hindered by certain factors.

In certain institutes, the percentage of middle-aged groups has decreased continuously. For example, this trend is observed at the Slovak institute of physics. During the transformation period, the institute has managed to increase its share of professionals holding scientific degrees aged 40 or younger. Meanwhile, researchers aged 41-50 have left the institute, thus creating a gap in this middle age group. This causes a problem for the change of generations in the institute. Who will be the successors of those who are due to retire in the near future? In 1990, there were 2.2 staff within the age-group 41-50 for every one staff member aged over 50. In 1992, that ratio went down to 1.3; and in 1994 to 0.7. Similar tendencies are apparent in other institutes. In the Romanian institute of physical chemistry, 30% of research personnel are younger than 30, 42% are above 45 and only 25% comprise the age group of 31-45. In the Czech institute of physics in 1989-1994, due to the efficient recruitment of young staff the share of employees under 40 remained constant (at 33%). Simultaneously, the oldest group rose from 33% to 42%, while the percentage share of the middle-aged group decreased from 34% to 25%.

Purely quantitative changes are often aggravated by the decreasing qualifications of personnel – due mainly to a long-term brain-drain of the best and brightest. This is a particular burden faced by many institutes in attempting to ensure a smooth succession of generations. Young scientists are needed both for research activity and for training as future leaders of scientific teams and units. But under the prevailing conditions, an evolutionary development of personnel potential cannot be expected. It seems more likely that the immanent retirement of many leaders from research and from the universities will have far-reaching consequences for the institutes studied. On the one hand, there will be a unique possibility to introduce drastic changes in research priorities and to terminate outdated programmes that have lost either their urgency or their practical importance. On the other hand, there is the danger of an uncontrollable break in the continuity of scientific traditions.

This is a particular danger for research projects which involve a small number of researchers for their future development, because the critical mass necessary for the well-being of the research activity can be easily lost. Even in such a huge country as Russia this danger is visible. For example, an expert from the Russian institute of economics is especially anxious about the situation: "The institute is doomed to a slow, natural death."

One of the most dangerous consequences of collective aging is that the existence of 'schools', or research teams, committed to a specific research programme is endangered (Mikulinski 1977, 115). These are teams which in the Soviet tradition usually consist of one leader and no more than some dozens of scientists. Many of the experts polled were pessimistic about the future survival of these teams. A respondent from the Russian institute of eonomics said: "We cannot attract young researchers. The institute will gradually die. I can't see any prospects." In the opinion of another interviewee from this institute, "though the scientific school created by its director still exists within the institute, there are no young people ready to acquire the knowledge and experience of the older generation. The continuity is broken and, obviously, will never be restored. If the institute, nevertheless, survives, it will become quite different."

Similar views were expressed at the Belarussian institute of engineering and at the Russian institute of biology, both of which are in a comparatively better situation with respect to their personnel structure than most others. But, even here, pessimism reigns as to the future prospects of maintaining the continuity of their research programmes. The destruction of teams seems set to become an acute negative consequence of the collective aging of scientific manpower.

The high fluctuation of personnel (primarily, of young researchers) creates an unstable situation. Commitments undertaken may be endangered because leading researchers can leave the institute at any moment. "I sign the contract and then I feel myself as if on a powder keg", complained a laboratory head at the institute of engineering in Belarus. This instability prevents the long-term development of priority research programmes.

It follows from individual aging in research activity that nature gives scientists very little time to realise their creative potential. Almost their entire research career depends on whether a young researcher aged between 20 to 30 manages to enter science, to pass through the adaptation phase, and then to start independent work. The initial conditions provided for researchers at 30 are a necessary prerequisite for the fullest realisation of their creative abilities at 40. A long-term decline in the social prestige of a scientific career and, hence, the loss of its appeal for young people means that at least one young generation of university graduates will be lost to fundamental science.

5. Coping Strategies and Tactics

The devastating consequences of collective aging in science can only be prevented by a strong personnel policy that does not attempt to retain the intellectual core but focuses on its renewal. However, such a policy is basically prevented by the financial situation of the academy institutes. This is perhaps most pronounced in Russia where interviewees from the institute of biology indicated that an atmosphere of passivity and despair reigns in the institute. In addition, in practice, control of collective aging is often minimal because of several interrelated reasons: the uncontrollable voluntary departure of researchers; the restrictions of labour legislation; the reluctance of young researchers to enter science; and the relatively poor qualifications of newly-hired young researchers.

Nonetheless, a more thorough analysis of the case studies shows that there are also some signs that inspire optimism. For instance in the Slovak institute of experimental pharmacology, the average collective age has become 5-10 years younger. Whereas in 1990, the age of laboratory heads varied between 40 and 60, in 1994 it ranged from 32 to 53. At the Lithuanian institute of biochemistry the personnel age structure is also quite favourable: 44% of the researchers are younger than 40 and only 20% are older than 50. In most of the Romanian institutes researchers younger than 30 constitute 20% and, in some of them, even 40-50% of total personnel (Radulescu 1995).

As a rule, these favourable developments are a result of the purposeful activities of the institutes' management. A typical step is the dismissal of all researchers who have reached retirement age. For example, this was done at the Slovak institute of experimental pharmacology and at the Polish institute of physics. 250 researchers were retired, and only young people were hired to replace them.

Another trend in the personnel policy of research institutes is the recruitment of people who are predominantly university graduates and postgraduates to fill vacant jobs. This was done by many managers of the institutions studied (e.g., the institutes of polymer mechanics and organic synthesis in Latvia and the institute of physics in Poland). Generally, these steps produced positive results. Conversely, attempts made to preserve a full complement of personnel by any and all means, aggravate the situation and lower the collective level of qualifications. For instance at the Ukrainian institute of informatics engineers from its 'own' design office filled the vacancies that arose after the dismissal of researchers. Respondents think that because of this, the institute has become a kind of diminished model of the applied industrial R&D complex.

The different kinds of factors that affect research personnel mobility interact and become actualised at the level of the researcher's personal decisions which motivate him or her either to leave or to remain.

What forces or seduces young people out of science? This question was answered in questionnaires distributed among researchers. For example, at the Lithuanian institute of biochemistry the reasons for leaving emerged as follows: low remuneration, 82%; no prospects of a research career, 32%; stressful team relationships, 25%; and inadequate research equipment, 21%. According to the respondents from the Slovak institute of physics, young people are reluctant to work in science because of changes in their values, the uncertainty of research policy and the low number of postgraduate fellowships. At the Romanian institute of theoretical informatics it is thought that inadequate equipment was an important reason for researchers leaving since, without up-to-date equipment, engagement in research to an international standard is not possible.

At the Czech institute of physics respondents similarly indicated the factors that discourage graduates from working there: firstly, the minimal attractions of a scientific career, especially its poor remuneration; secondly, the unsatisfactory conditions for engaging in professional activity – especially due to outdated experimental equipment – as young people only express their potential interest in well-equipped laboratories or in theoretical research; and, finally, because the pool of physics graduates is small. It is evident that there are both material and immaterial factors at play here.

The flip side of the coin is to ask: what keeps people in science? The questionnaires distributed at the Russian institute of biology have shown that for 73% of the respondents the main 'anchor' is that the work is interesting; 37% express unwillingness or an inability to change their way of life; and 32% hope for a change for the better. For the respondents from the institute of physics in Belarus, among the factors that keep researchers in science, confidence that the crisis will be overcome (47%) and the availability of conditions for creative self-realisation (21%) rank the highest. For many young people, preparation of a doctoral dissertation is an important stimulus. A respondent, aged 32, from the Romanian institute of biochemistry, answering the question "Why do young people come to the institute?", wrote: "I came to the institute to create something new, and it is only possible at a research institute. At present, my salary is low, but in the future I will do my best to have excellent results and my income will be higher." Thus, a science policy addressed at attracting young people should be based on balancing the material and immaterial incentives.

In the institutes examined, respondents expressed their opinion about recruiting young people to science and the most prevalent suggestion was that salaries should be substantially upgraded. In Russia this opinion was expressed by 92% of the respondents at the institute of economics and by 83% at the institute of radio engineering and electronics. A stable inflow of young researchers in the Romanian institute of theoretical informatics was mainly achieved by offering creative kinds of work coupled with safe employment positions, financed from the institute's budget. However, respondents are rather pessimistic in their forecasts: they judge that young professionals, once married, will probably leave the institute for more profitable sources of income.

Of course, solving this problem is far beyond the scope of any one individual institute. In most of the transformation countries, amongst the branches of the labour market, science has become secondary in terms of salary rates. In particular, in Belarus the average monthly salary in the institutes of the Academy of Sciences was 92% of the national average wage in 1992, and 90.1% in 1995. Young researchers leave the Eastonian institute of chemical physics for private businesses which command salaries of 5,000-10,000 Estonian crowns as against the 1,500-2,000 they may earn at the institute. This tremendous difference makes the academy institutes absolutely unable to compete with the more prestigious branches of the labour market.

Given these differences in salaries, the institutes have to look for ingenious ways to compensate for poor remuneration. Diverse means of offering young researchers alternative forms of material support include, for example, compensating them for losses incurred as a consequence of their undertaking postgraduate work as compared to those researchers who prepare their doctoral dissertations while holding down a full-time job (institute of physics in Poland). At the Latvian institute of organic synthesis students are hired, and paid, for their practical work at the Institute. At the Russian institute of radio engineering and electronics young researchers – unlike others – are not transferred to part-time work regimes and they also receive a special additional remuneration from the director's fund.

A grant-based financing system can be a great resource for offering young researchers material support. Scientific foundations arrange special competitions for young researchers, as is done by the Belarussian Basic Research Foundation, for example. Special criteria concerning research teams have been added to the usual criteria for research competition by the International Soros Science Foundation (International Science Foundation 1995). At the institute of biochemistry in Romania either a foreign Doctorate fellowship, or research grants under PECO, TEMPUS and other programmes are guaranteed to young scientists for periods ranging from three months to a year.

Unfortunately, because of funding shortages, official measures for supporting young researchers are often insufficient. For instance, in Belarus, prizes and staff fellowships are available to young researchers, but the actual money they offer is so low that they have become a moral rather than a material stimulus. The Hungarian parliament annually allocates 150,000-160,000 Ft to newly-hired young researchers payable over a three-year period. The institute of isotopes allocates 1,000,000 Ft. to every department that recruits a researcher younger than 35. However, respondents admit that these activities hardly foster personnel rejuvenation.

International scientific co-operation is extremely important for recruiting and 'anchoring' young people. It combines material and moral incentives, higher personal incomes for researchers and better conditions for professional activity. The wish to go to the West for training and work opportunities is widespread among young people in all the post-socialist countries (see Mirskaya, this volume). This desire is utilised effectively by most of the institutes studied within their personnel policies. Therefore, international co-operation is now deemed to have become extremely important for the continued well-being of science in the transitional countries (Schweitzer 1995).

The institute of biology in Poland gained experience of personnel rejuvenation with the help of a particular innovation in international co-operation. In 1990 the so-called 'Trustees Group' was formed as an initiative of the institute. This group is made up of persons who have left the institute and now hold independent positions in Western laboratories. Close co-operation with them is maintained. They give guest lectures in the institute, look after young scientists on internships abroad (many of which are postdoctoral), and participate in joint research projects. During the period of martial law, around 30 scientists did not return from their foreign travels. Now the trustee system is becoming the nucleus of a programme to encourage young researchers to return to Poland from abroad. The regulations for using the scholarship fund to cover the costs of researchers returning from abroad – called the Fund for Returns – were approved in July 1992. This fund is an integral part of the institute's scholarship fund. It is intended to cover the costs of recruiting persons returning to the institute from long-term internships abroad or from other long periods of employment abroad. The benefits from these funds are continued for a period of three years following the person's return. Eight to ten people have returned to the institute as a consequence of this initiative – most of them young. The programme has been so successful that it is now highly likely that young people who leave for temporary internships will return. Choosing to remain abroad has become the exception rather than the norm. The main arguments in favour of returning to the institute of biology are that returnees enjoy greater scientific independence than they would in the West; that they face less competition; that they can depend on greater employment stability and security; and that they are offered greater chances for self-fulfilment. Returnees have the opportunity to work in their own well-equipped laboratories, a chance which would not be so quickly offered to them in the West.

In most of the institutes studied, respondents stress the important role that a good material and technical base plays in retaining research staff. After all, more than for any other profession, the working conditions for research – the equipment, the possibilities of obtaining information, etc. – represent more than mere tools of the job. Good working conditions provide an essential motivational basis for staying in the field (Radulescu 1995).

Many respondents have emphasised the importance of training as an essential component of policies addressed at recruiting and retaining young research staff. Elements of such policies include the establishment of university departments (institute of isotopes in Hungary); supporting the professional orientation and selection of young scholars by means of organising Olympiads, contests, and summer and winter schools (institute of biology in Russia); recruiting undergraduates and graduates from Eastern Europe, particularly from the Ukraine and Yugoslavia (institute of physics, Czech Republic and the institute of isotopes, Hungary); and searching out foreign fellowships for training young researchers in the West for periods lasting from six months to three years (institute of experimental pharmacology, Slovak Republic). This is to name only a few of the activities organised by academy institutes for training young researchers.

Policies supporting young research personnel have become highly individualised. A respondent from the Russian institute of biology, in whose laboratory only one young researcher continues to work, noted that this researcher occupies a special position as compared with the other team members: "We give him everything we can. Now he receives a higher salary than any of us. For example, I haven't travelled anywhere for more than four years. For him, it is more important. We cannot let him 'grow potatoes', he must work. All the time. But he is young, has a young family. We cannot let him leave us, leave the country. We have no future without him. Neither does our IDEA. More young people are wanted. Urgently wanted. So much can be done!"

At the institutional level, this reflects the commitment to support young researchers in order to ensure the continued development of scientific ideas. Obviously, local responses such as this one cannot replace a systematic and sustained personnel policy at the national level which must ensure adequate salaries, a training and refreshment training system for researchers and access to international co-operative research activity. But the instance cited above attests to the fact that at the core of science we find the professional nurturing of the younger generation, beginning with the selection of talented students from secondary school and finishing with the supervision of doctoral dissertations, as it is in the institute of physics in the Czech Republic.

The experience acquired by the institute of physics in Poland suggests that locally-based research personnel policies hold considerable potential. After the initiation of reforms, new personnel policy instruments were introduced in order to stimulate more effective work and to compensate for losses incurred by studying for a doctor's degree (this has already been discussed in the description of the training functions of the institute). In 1992 a preferential system of wage rises for research workers based on an assessment scheme was introduced. Preferences were given to:

- young employees enrolled in extra-mural doctoral studies for five years (with the possibility of an extension for one year) and interns over a preliminary period of not longer than two years; and,
- scientists who earned their doctorate within four years, for a period of one year after being awarded the degree.

It is evident that this institute does its best to recruit young researchers. The employees of the institute not only lecture on their specialist subjects in universities but also teach popular science in secondary schools, promote scientific Olympiads, etc. If the institute still encounters recruitment problems, they are caused by the declining popularity of science amongst university students. This stands in contrast to the institute of biology in Poland, for example, which has fewer recruitment problems because its field is currently a fashionable one which arouses great interest amongst students. It is obvious that this aspect of public opinion – while an important determinant of an individual's career choice – is a factor which lies beyond the control of any given institute in its quest to attract young candidates to its own particular field. Nor is it a problem specific to the ex-socialist countries, as the case of the institute of chemistry in the Czech Republic indicates. The image of chemistry as a destructive force which damages the environment is as poor in Eastern countries as it is in the West. However, the wheel may turn again.

6. Conclusion

The case study findings on aging and on the age structure of researchers in the academies' institutes – even allowing their non-representative nature – clearly indicate how dramatic the current situation is, as well as the threat this issue poses to the future of science in the countries of Central and Eastern Europe. Any serious imbalance in the age structure of researchers must ultimately result in an uneven succession of generations and, consequently, must affect the innovative process which that succession sustains. The focus on age – and therefore the mechanism of intergenerational succession – reveals that the problem for these countries is not simply one of finances. Beyond the crucial issue of funding lies the much more intractable problems of the economic and intellectual environments which affect career choices and the calculations that drive them. Changing these is not easily accomplished, nor can they be redirected over a short period of time. The crisis of the transformation period in the Central and Eastern European countries, insofar as the intergenerational succession of researchers is concerned, will be felt for years to come.

References

Analytical Centre for Science and Industry Policy (1996). *Mobility of scientific personnel in Russia.* Moscow: Analytical Centre for Science and Industry Policy (in Russian).

International Science Foundation (1995). *1994 Annual Re*port. Washington, DC: ISF.

Kuhn, Thomas (1970). *The structure of scientific revolutions,* 2nd ed.. Chicago, IL: University of Chicago Press.

Lehmann, N. C. (1953). *Age and achievement.* Princeton, NJ: University of Princeton Press.

Levin, Stephan, Paula Levin, & Sharon Levin (1993). Age and the Nobel prize revisited. *Scientometrics* **28**, 387-399.

Messeri, Peter (1988). Age differences in the reception of new scientific theories: The case of plate tectonics. *Social Studies of Science* **18**, 101-121.

Mikulinski, S. (ed.), (1977). *Scientific schools.* Moscow: Nauka (in Russian).

Nesvetailov, G. A. (1986). *Intensification of academic science.* Minsk: Nauka i technika (in Russian).

Pelz, Donald C., & Frank M. Andrews (1996). *Scientists in Organizations. Productive Climates for Research and Development.* New York: Wiley.

Radulescu, Dan (1995). Romanian youth and scientific research. Academies in transformation. *Higher Education in Europe* **20,** 220-229.

Rappa, M., & K. Debackere (1993). Youth and scientific innovation; the role of young scientists in the development of a new field. *Minerva,* **1**, 1-20.

Schweitzer, G. E. (1995). Can research and development recover in Russia? *Technology in Society* **17**, 121-142.

Simonton, D. K. (1988). Age and outstanding achievement: what do we know after a century of research? *Psychological Bulletin* **104**, 251-267.

Stankiewicz, R. (1979). The size and age of Swedish academic research groups and their scientific performance. In UNESCO (ed.), *Scientific productivity. The effectiveness of research groups in six countries* (pp. 191-222). Paris: UNESCO/Cambridge University Press.

Stewart, John A. (1986). Drifting continents and colliding paradigms: a quantitative application of the interests perspective. *Social Studies of Science* **16**, 261-279.

UNESCO (ed.), (1979). *Scientific productivity. The effectiveness of research groups in six countries.* Paris: UNESCO/Cambridge University Press.

Zuckerman, H., & Robert K. Merton (1972). Age, aging structure in science. In M. W. Riley, M. Johnson, & A. Foner (eds.), *Aging and society, Vol. 3: A sociology of age stratification* (pp. 292-356). New York: Russell Sage Foundation.

FROM EAST TO WEST: NEW PATTERNS OF INTERNATIONAL RELATIONS OF RESEARCH

ELENA MIRSKAYA

Institute of the History of Science and Technology Russian Academy of Sciences 103 012 Moscow

1. A Look at the Past

In ex-socialist countries the issue of international collaboration attracts much attention and raises high emotions. International collaboration had always been an aspect of scientific life where a scientist's dependence on the party and administrative authorities had manifested itself in a most evident and offensive form, although there were differences between the various socialist countries.

In the Soviet Union, international collaboration was long perceived by scientists as a lucky, but practically unavailable, opportunity to go abroad for several days. Everybody aspired to "go abroad": natural scientists and scholars pursuing humanities, prominent and middle-level scientists, men and women, the young and the aged. Everybody dreamt of travelling into an unknown world, into an international community of colleagues, of meeting men of legend, demonstrating one's worth in their presence, of buying foreign gifts for the whole family, and on coming back, of telling stories about "the mythical life abroad" for months.

Over time, trips to COMECON countries ceased to be a rare and exotic thing. Every research field had its more or less established scientific contacts and some on-going scientific collaboration even though, sadly enough, the latter was sometimes largely ritual and inefficient. Nevertheless, until the COMECON system was dismantled every trip to any country whatever was regarded as desirable and prestige-bearing. The quota of foreign trips allocated to an institution by the Academy Presidium was distributed by the institute heads together with a special committee for international collaboration on the basis of

R. Mayntz et al. (eds.), East European Academies in Transition, 107–123.

the references presented by the local office of the Communist Party.[1] To avoid collisions and conflicts, a great number of criteria were elaborated with a view to providing an unbiased distribution of foreign trips. Very often, however, non-scientific and unpublicised arrangements and motivations were, in fact, decisive.

Travelling to "real" (capitalist) foreign countries was simply a miracle for the average scientist.[2] This opportunity was available only to top scientific administrators and to a very few trusted elite scholars. For all other researchers, personal invitations received from foreign research institutions impeded, rather than facilitated, the opportunity to go abroad. Publications in international and foreign journals which determine the scientist's reputation in the world of science or co-authorship with foreign colleagues were rare phenomena (Science & Engineering Indicators 1987, 286f.).[3]

Scientific co-operation within the COMECON framework certainly had its good points. The feeling of equality, a certain human openness in scientific contacts, and a lack of competition are sometimes recalled today with nostalgia. Of the three basic values *liberté, egalité et fraternité*, the last two were realised, while the absence of freedom united the men and women of science perhaps more than its presence would have done. Although genuine scientific exchange was extremely hampered, contacts within the COMECON framework received unfailing financial support from the state, which attached considerable importance to all the rituals of scientific collaboration that served to demonstrate a high level of socialist science and a close integration of the COMECON countries.

The situation in the other socialist countries was not exactly the same as that in the Soviet Union. Some of the countries in Central and Eastern Europe (CEE) never experienced strict isolation from Western contacts (Science & Engineering Indicator 1987, 285).[4] This made a considerable difference when, in 1990, the COMECON ceased to exist. In most CEE countries the search for, and development of, scientific contacts with the West had started much earlier. During the transformation period of the 1990s, international collaboration de-

---

1 Since visits abroad largely followed the rule of equivalent interacademy exchange, the chances of foreign travel were markedly lower in the (large) USSR Academy of Sciences than in the academies of smaller countries.

2 In the Soviet republics the situation was even worse, since Moscow had monopolised all the mechanisms of international contacts.

3 In the mid-1980s the share of papers with a foreign co-authorship amounted to 9.7 % in the highly egocentric US, while in the Soviet Union it was 2.5%, including all COMECON relationships. Actually for the USSR this proportion was even lower since the cited figures are based on the *Science Citation Index* which scans only high-rank Soviet journals.

4 Thus till the mid-1980s the number of Polish scientists visiting the USA in a given year was from five to ten times as high as the number of Soviet scientists doing so. Some Hungarian scientists have also long since found their place in the international scientific community.

pended in many academy institutes on international contacts previously established. In the USSR, *Perestroika* and the weakening of the Soviet regime which set in during the second half of the 1980s promoted relationships with Western science here as well. However, with the break-up of the Soviet Union, its ex-republics generally found themselves in quite unequal positions, which also held true for the level of their international science relations. Against the background of the economic crises that soon overwhelmed the post-communist countries, all international relationships, contacts and interactions gained a different significance, becoming important for professional survival. In the new situation, international co-operation became a factor of particular importance.

2. A Note on Methods

The case study material which forms the basis of this book does not permit a systematic and representative overview of international collaboration, as the frequency of international contacts and the intensity of international collaboration vary considerably between different institutes in any given country. For one thing, these differences reflect the (variable) international reputation of the institutes. For another, the pre-transformation intensity and pattern of external contacts has importantly shaped subsequent developments. A third important factor is the attitude of the institute head toward international collaboration, and his/her skill in opening up avenues towards it. These factors cross-cut the variance between countries which stems both from the pattern of their pre-transformation relations to the international science community and from differences in the legal, bureaucratic and financial conditions presently impinging upon them.

To broaden the data base, and because the relations of CEE scientists to the international scientific community may be expected to feed back into the process of political and economic transformation, the topic of international relations in science has also become the object of a separate, depth-oriented survey that was added to the case studies conducted in 1995 (Mirskaya 1995a).[5] This survey was only realised in Russia and had been preceded by a pilot study carried out in 1994, which analysed the state of research and researchers in the Russian Academy of Sciences (Mirskaya 1994, 1995b). The survey attempted to find out the latent correlations between scientists' participation in international co-operation and other aspects of their professional life. It covered eight institutes of the Royal Academy of Sciences (RAS) in Moscow: two physics institutes, three chemical institutes, two biological institutes, and an institute for social

5 Research covering this topic was also supported by grants of the Russian Foundations for Basic Research (Nr. 96-06-80260) and for the Humanities (Nr. 96-03-04599).

science and humanities (see International Science Foundation 1994, 9).[6] The sample drawn from these institutes[7] was a non-representative target sample, formed by way of a purposeful selection of sectors and laboratories which had, in fact, established international contacts. This proved necessary in view of the fact that in a representative sample, the incidence of relevant international relations would have been too small to permit statistical analysis using the intensity of such relations as a major variable. The study was conducted by way of interviews, using a single unified questionnaire. In addition, eight in-depth interviews were carried out with institute directors and vice-directors. Taking into consideration the peculiarities of Russian science as heir and successor to the gigantic Soviet science system, the patterns found here cannot be regarded as representative of those found elsewhere in Eastern Europe. But the survey data still give some insight into micro-level aspects of international co-operation that may well hold for other CEE countries as well.

3. Changes in the Geography of International Contacts

Among the changes that occurred in the international scientific collaborations of ex-socialist countries in the 1990s, the most evident is a radical transformation of the geography of contacts, which manifested itself in an abrupt redistribution of the proportion of "Eastern" and "Western" international contacts.[8] The dissatisfaction with the poor efficiency of scientific collaboration within COMECON that had accumulated over time, the sudden shift to a higher level of autonomy and, above all, the sharp reduction of science funding in all the newly-independent countries together acted to promote an orientation towards Western contacts and to give them priority in the pattern of international relations. However, it was not always easy to realise these new aspirations. While before the transformation the expenditures for science – including the costs of international scientific contacts – were provided, directly or indirectly, by the state, the international contacts of East European scientists who had obtained political freedom were now found to be financially utterly dependent on Western partners. Soon, many scientists came to regard relations with Western partners or institutions as a source of funding, ensuring not only international com-

6 These were the Lebedev Institute of Physics, the Shubkin Institute of Cristallography, the Kurnakov Institute of General and Inorganic Chemistry, the Institute of Physical Chemistry, the Zelinskii Institute of Organic Chemistry, the Kol'tsov Institute of Development Biology, the N. Vavilov Institute of General Genetics, and the S. Vavilov Institute for the History of Science and Technology. – In the ISF list of fifty top institutions, the surveyed physical institutes rank 4th and 29th, while the chemical institutes rank 13th, 32d and 40th.

7 There were 302 respondents: 92 from the physical institutes; 120 from the chemical ones, 70 from the biological institutes, and 38 from the social science institute.

8 The notion "Western" usually embraces all advanced countries including Japan.

munication but also providing an opportunity to continue research work. Western contacts, in other words, helped in the struggle for professional survival.

A positive aspect of the political transformation in the CEE countries was the liberalisation (legal or de facto) of the conditions – the requirements and restrictions – of international collaboration of all types and on all levels – from ministries to institutes and to individual scientists. The fact, as well as the positive impact, of the greater liberalism of the rules governing international scientific connections was reported by all network participants without exception. This reduced the necessity for bureaucratic mediation and made it possible for lower-level actors to enter into contracts. Thus, Polish academy institutes, though not legally independent entities, began to enter directly into contractual relations with foreign scientific partners. The institute of physics of the Polish Academy of Sciences, where in 1990 only three of the 25 projects conducted on the basis of bilateral agreements with (nine) foreign countries were based on direct contracts, collaborated in 1993 with 37 of a total of 60 institutes (in 23 countries) directly, without official agreements between higher authorities. Western institutes and institutions represented the majority among the new contacts. However, collaboration with former partners from the socialist block, even though weakened, still continued.

The preservation of scientific contacts within the former COMECON area was mentioned (mainly with respect to Russian research institutions) by the Polish, Bulgarian and Hungarian case studies, particularly in the field of physics. Of the 60 physicists from the Bulgarian institute of nuclear research who go abroad each year to carry out experiments, 40 go to Western centres and 20 to Dubna in Russia. Slovak respondents indicated that the number of Eastern visitors has decreased, but their rank (research level) now tends to be higher. Romanian respondents pointed to a more widespread problem when they referred to a lack of real opportunities for establishing productive communication between partners who all contend with insufficient funding (see Nauchno-organizatsionnaya deyatelnost'v 1994, 93).[9]

The state of international relationships in the former republics of the Soviet Union was essentially different from other CEE countries when the former regime broke down. The co-operation of the former Soviet republics with Russia, and with one another, was a habitual, long-established interaction within a single whole; at the same time, scientific relationships with other, especially Western, countries was largely routed through Moscow. Thus, while scientists of

9 It is pertinent to note that in 1993–1994, the RAS, seeking to restore and strengthen its relationshipswith the countries of Central and Eastern Europe, signed new agreements on interacademy exchange with the academies of Bulgaria, Hungary, Poland, Romania, the Slovak and Czech Republics. In 1994, the greatest amount of scientific exchange was realised with the Academy of Sciences of the Czech Republic (76 man-weeks), the Polish Academy of Sciences (72 man-weeks), and the Hungarian Academy of Sciences (60 man-weeks).

East European countries (except Romania) expanded their previously-existing Western contacts in the period of transformation, the scientific communities of the former soviet republics were suddenly divorced from their centre and thus were largely forced to initiate contacts from scratch.[10]

The somewhat unstandardised nature of the data reported in the case-studies does not permit us to indicate the priority partners for each of the participating countries. In general, Austria, Belgium, Canada, China, France, Germany, Great Britain, Italy, Japan the Slovak Republic, the Republic of South Africa, Spain, Sweden, Switzerland and the USA were cited.[11] This list is compiled in alphabetic order; the available data do not permit an estimate of the extent, or the significance, of international relationships with different countries. Also, the notion of "contacts" often remains undefined and can cover quite different forms of interaction – from the establishment of joint laboratories (e.g., Poland with France) to private talks between individual scientists about possible future research co-operation. Nonetheless one can note that the former socialist countries apparently tend toward intra-European collaboration, chiefly with Germany. But there are exceptions to this tendency. Russian scientists have extensive contacts with the United States. In biological research, collaboration with American scientists generally plays an important role. The recent interest in scientific contacts with China (reported by Belarus, Poland and Russia) is also worthy of mention.

The quantitative, as well as geographical, expansion of international contacts was not only due to a changed political and legal framework. An important motive force was the personal initiatives of the most active scientists. In fact, an absolute majority of the project participants in all the countries emphasised that the development of international contacts is the result of personal initiatives of individual scientists who act either individually or as leaders of research groups. This also came out very clearly in the Russian survey, which found that, in the period covered, international contacts developed largely at the level of research groups and individual scientists, but played a minor role in the overall functioning of academy institutes.

However, there seems to be a tendency to underrate the significance of institute heads for the development of international collaboration.[12] The decisive

10 It must be borne in mind that even though the formally recorded "onset" of the transformation process varies only insignificantly for thecountries under comparison, this does not hold for the duration and present state of the process. In some countries (e.g., Hungary) a new system of international co-operation has long been established, while in others (e.g., Romania or Belarus) it is still in the initial phase.

11 Countries participating in the network project are not included in the list.

12 That the administrative effort of the institute heads is a significant factor was reported only by Romanian respondents; the importance of the scientific authority and of the personal position of the institute leaders was emphasised also by Belarussian and Czech scientists. The

role of administrators and the power of hierarchy certainly belong to the past,[13] but their involvement is still necessary if the newly-emerging patterns of contact are to be stabilised. A largely decentralised pattern of establishing foreign contacts cannot be considered an unequivocally positive feature. Collaborative efforts carried out by individuals or a team are highly motivated and efficient, but in the long term this type of connection has its drawbacks: in *atomising* research, it induces a certain amount of chaos in the system. In the initial phase of transformation, and against the background of a history of planned research on topics which continued unchanged for decades, the decentralised pattern of collaboration may have a positive impact and induce productive changes. However, if not complemented by institutionalised forms, this kind of collaboration supports only the most advanced scientists and research isomorphic to Western ones.

4. Forms of International Co-operation

In the early 1990s, the forms, as well as the geography, of international interactions underwent important changes. It is in this period that innovations which bear testimony to an essentially new state of international relationships have evolved and spread. This process was promoted by a fortunate combination of internal and external factors: the willingness of advanced countries to support research in the CEE countries and to help their scientists to integrate into the international scientific community, coincided with the liberalisation of the initiative of individual scientists and research groups from political interdictions and bureaucratic restrictions.

Three basically different types of international collaboration in science can be observed in the CEE countries: (1) support rendered by foreign philanthropic or scientific foundations awarding research grants to be spent at the home institution; (2) research collaboration between institutes or individual scientists/ research groups; and (3) indirect participation of CEE scientists in the international scientific community by publication in foreign journals or on the basis of computer telecommunications (e-mail, etc.).

International relationships involving direct contact with foreign partners, and thus falling largely into the second category, can range from single trips undertaken by East European scientists with a view to attending international meetings, improving their proficiency and/or carrying out some experiment, to the establishment of joint research centres and laboratories. In the case studies,

---

dependence of successful international interactions on a correct scientific strategy implemented by the institute heads was indicated only in the Latvian and Polish studies.

13 Only in the Slovak case studies was it mentioned that an application for joint research or a grant requires approval by the institute's scientific board and heads.

publications in international journals, co-authorship with foreign colleagues, foreign visits, and the active use of electronic communications have been cited as essential indicators of a scientist's or institute's incorporation into the international community. But the most important forms of international interaction are those that represent authentic scientific collaboration in which both parties can mutually realise their interests. This is true in the case of long-term research carried out within the framework of intergovernmental and interacademy agreements, as well as in the case of joint research carried out under direct contracts between scientific institutions and involving direct co-operation at the level of research groups and individual scientists, leading to the establishment of true international research collectives. The most impressive examples of a real integration of Eastern and Western research systems are institutionalised forms of long-term international research collaboration, as for instance in the Polish-French centre for biotechnology, the American branch of the RAS institute of theoretical physics, or the Belarussian–German laboratory for information technology.

Target grants allocated by advanced countries and special programmes aimed at supporting CEE researchers and research in the difficult transformation period were an essentially new phenomenon of the 1990s, when science in the countries undergoing reform was in a critical financial situation. The activity of foreign philanthropic foundations, specifically the G. Soros Foundation which awarded a considerable number of rather substantial grants for research, travel and electronic communication, was of great importance to scientists.

Though the different forms of international relations have not developed in exactly the same way in the different countries, or even within different institutes in a given country, some common features still emerge from the case studies. Thus, travelling abroad is generally the most popular form of international contacts. An unprecedented growth of *opportunities* to attend international scientific meetings is mentioned in all the countries. However, the radical increase in opportunities which are due to abolished restrictions does not easily translate into real potentialities because of the lack of funds. "The opportunities are brilliant, only funds are lacking" – this is the most widespread remark of respondents in the case studies.[14] The de facto involvement of researchers in international co-operation is, of course, not uniform so that the amount of foreign travel varies greatly between individual countries and institutes; in some cases the difference amounts to 1:10.

Thus in 1994, the Czech Republic's institute of physics reported 520 scientific trips, while in the Polish institute of biology with quite well-established scientific collaborations 75 scientists travelled abroad; in the Slovak Republic's

14 The total financial dependence on foreign partners and organisations in this respect is most painfully felt by the scientists of the former Soviet Union who previously enjoyed state support but who now experience difficulties.

institute of physics, the number of foreign trips increased from 81 to 162 between 1990 and 1994. In the Belarussian institute of physics the amount of travelling abroad also doubled in the years of transformation, but only from seven to 14 trips, while the institute of engineering reported nearly a 20-fold increase (from two to 39). In Romania, where attending international meetings was practically banned till 1989, 35-40 associates of the institute of physical chemistry went abroad on research missions annually (with expenditures borne by the institute), and 30 of the 32 associates of the institute of biochemistry spent several months in foreign research laboratories. A decrease of contacts involving foreign travel during recent years has been reported by Bulgarian scientists; here a drastic deficit of centralised national funding – which is a common problem – was aggravated by the prohibition to spend the money earned through extra-budgetary contracts on foreign travel.

In some of the countries, some institutes particularly mentioned foreign travel for the purpose of training.

Thus the Polish institute of physics arranges for three-month post-doctoral studies in recognised research centres with a relevant specialisation for its members. Each year, 20-30 scientists of the Hungarian institute of chemistry go abroad for advanced studies. All of the investigated institutes of the Romanian Academy of Sciences use every opportunity to send their post-graduates and post-doctorates to scientifically advanced countries (the institute of physical chemistry – 25, the institute of biochemistry – 20, the institute of theoretical informatics – 16); the post-doctorates can even advance their career while retaining 25-50 % of their salary during their absence, which is believed to stimulate their return. This form of collaboration occurs less frequently in the former Soviet republics, except for the Baltic countries.

In some scientific fields in the countries undergoing transformation the chance to work abroad to conduct empirical research that requires advanced equipment and powerful computers is of great interest. This practice does in fact exist and sometimes becomes a stable arrangement. Thus Polish biologists regularly go to France to perform their experiments there; Bulgarian nuclear physicists go to Trieste and Dubna; while Lithuanian scientists perform extramural experiments with new-generation equipment which their institutes cannot afford to buy. The researchers who have such opportunities achieve a level of proficiency in experimental science which they could not have gained within the previous system of scientific relationships.

A considerable number of researchers from the former socialist countries work in various foreign research institutions under long-term contracts of different duration. This type of activity is not necessarily a form of international collaboration; often it is just work-for-hire when the skilled labour of a scientist-*gastarbeiter* is employed, while the scientist himself gets nothing but money out of it. In other cases a long-term incorporation into advanced foreign research teams can essentially improve a researcher's professional knowledge and skills and impacts positively on his/her future work.

Four or five scientists annually have left the Hungarian Academy of Science (HAS) institute of isotopes for a period of over six months in order to work abroad under contract; the general trend reported for Hungary was a gradual decrease in the number of advancement studies and an increase in such contract work. In contrast, Bulgarian biologists reported an abrupt decrease of this type of activity (34 scientists in 1992, but only three in 1995), while nuclear physicists estimated the number of their colleagues having foreign work contracts at 20-50. The Lithuanian institute of biochemistry reported 29 researchers working abroad, the Estonian institute of chemistry reported 20-25 each year. Many Russian researchers work abroad on a contract basis, chiefly in the USA, but it is quite difficult to distinguish between those who do so temporarily and those who have practically emigrated.

One cannot say that the scientists' visits are in one direction only, but the situation of the 1990s precludes a "symmetric" exchange, so that movement to the West is much greater than any counter-traffic. The great interest in ex-socialist countries that suddenly arose among Western scientists in the late 1980s and that led to a mass "pilgrimage" to the East, especially to Russian research centres of high prestige, died away very soon. Only institutes with well-established genuine collaboration and institutes that organise international meetings could maintain previous levels of foreign visitors.

The Polish institute of biology (72 visits in 1994) and the Czech institute of physics (128 visits in 1994) exemplify the first alternative, while the Hungarian institute of chemistry that organises meetings each year and can welcome 180–250 visitors annually is a vivid example of the second kind. Seeking to establish active international contacts, institutes of the Romanian and Latvian Academies of Sciences regularly arranged various research meetings; such meetings can also serve as a bridge between the countries of the former Soviet Union and the West. In general, however (except for Poland), the number of foreign visitors to Eastern institutes is two or four times less than the number of Eastern scientists going to Western research institutions.

Since the use of such terms as "joint project", "joint research" and "international programme" varies greatly between different countries, and many case studies report no specific data on this issue, it is rather difficult to make general and comparative statements about this type of collaboration.

Reliable information is only available with respect to 11 programmes in the Polish-French centre already mentioned; the participation of the Czech institutes of physics and chemistry in 48 and 6 projects, respectively; the increased number of contracts with foreign partners in the Bulgarian institutes of nuclear research and of molecular biology (5–9–7 and 1–1–6 contracts in 1992, 1993, and 1994, respectively); and three projects involving Slovak pharmacologists.

Of all the European Community (EC) programmes arranged to support scientists of the ex-socialist countries, respondents especially mentioned contacts with EUREKA, TEMPUS, PHARE, PECO, COPERNICUS, and TESSA. Though according to their aim and scale of funding, these programmes initially served to promote scientific communication rather than research, a number of scientists (Hungarian and Czech physicists, Bulgarian computer scientists, Slovak ecologists) were nevertheless awarded targeted research grants under the EC programmes as well as through American National Science Foundation (NSF) grants. In the Slav republics of the ex-Soviet Union, the best-known grants were those awarded by the G. Soros Foundation and later also by INTAS.

As reported in many of the case studies, especially from those countries which only recently have become open to international collaboration, the funds coming to the institutes through international channels accounted for an insignificant part of financing (only 2-5%) and could not markedly promote the functioning of scientific institutions.[15] However, for individual scientists and research teams engaged in joint projects and having foreign research grants, the resulting financial support is highly important and sometimes plays a decisive role in their professional activities. In spite of this, the Russian survey has shown that the majority of scientific personnel are dissatisfied with the level and scale of co-operation, which has manifested itself in repeatedly-cited intentions and efforts aimed at the search for partners and the elaboration of proposals for international projects.

This general impression is confirmed by data reported in the Lithuanian case-study report. Lithuanian scientists were asked to make up two rank orders: the first was to list, in the order of decreasing importance, the forms of international co-operation in which they were actually involved, the second was to list the preferred forms of international co-operation (in order of priority). As to *desirable* forms, the respondents gave first place to grants, second place to working within international programmes, followed by publications, attending international meetings, and professional training. But the rank order of *realised* interactions was nearly the opposite of this: attending international meetings (1), grants (2), professional training (3), and participation in international programmes (4).

Publications in international journals or editions, which are the alpha and omega of scientific activity, play a particular role in the integration of national scientists and institutes into international science. In some cases, publications are just a "prelude" – a first step toward international reputation which can promote the author's subsequent entrance into the sphere of international collaboration. In other cases, publications reflect the author's on-going involvement in international research and present the results of co-operative work. Only this second type of publication actually reflects international co-operation in re-

15 Only in the Latvian institutes did foreign contracts account for 16-17% of funding.

search, but the first type, too, indicates a certain form of integration into the international scientific community. Different proportions of these two types of foreign publications in the different countries could indicate different intensities of integration into the international scientific community, but the case studies have only been able to differentiate foreign publications according to their total amount.

In the Czech institute of chemistry, the share of publications in foreign journals amounted to 70% of all registered publications, in the Hungarian institute of chemistry to 50%, and in the Polish institute of physics to 30%.

Since 1990, the organisational possibilities for the scientists of most CEE countries to publish their work abroad have remained practically the same, although the aspiration to publish in international or foreign journals has increased greatly.[16] In Romania, where publications abroad were banned until 1989, and in the former republics of the Soviet Union where this procedure was absurdly difficult,[17] the opportunities improved radically, but their realisation required some time.

The Romanian institute of physical chemistry has reported very interesting data, comparing the annual number of published articles with the number submitted for publication. From 1990 through 1995 these figures were: 5/40, 18/42, 18/44, 32/56, 39/57. The Baltic case-studies have indicated a drastic change in the distribution of publications by language. The figures reported for Estonian publications issuing from academy institutes between 1991 and 1994 illustrate this interesting phenomenon: during this period, the proportion of books published in Russian constantly decreased from 16% in 1991, 14% in 1992, 12% in 1993, to 3% in 1994; for articles the figures are 25%,14%, 10% and 4%. The proportion of publications in English increased at the same time: 30%, 46%, 42% and 32% for books, and 44%, 55%, 62% and 7% for articles.[18]

On the whole one must bear in mind that the increasing number of publications abroad is not so much due to a general increase in the number of publications as to the reduction in the amount of national publications. But enthusiastic statements such as that "now anyone can publish his works in the West"- especially typical of the respondents from the Baltic states – emphasise only the legal aspect of the issue. The criteria by which submitted manuscripts are selected for publication in leading international journals are rather strict. Scientists

16 In some countries, there are also real incentives: in Poland foreign publications entail extra-payments; in Hungary, only publications abroad are taken into account in the evaluation of applications for national grants.

17 Sending an article not previously published in a Soviet source to a foreign journal required the judgement of an expert committee stating that ***the article contained nothing new and interesting.***

18 The general quantity of scientific publications diminished in this period. For instance, the annual number of books and brochures was decreasing from 103 (1991) – 72 (1993) – 73 (1993) – 68 (1994).

from ex-socialist countries are not generally accustomed to the fact that besides the permission, or legal possibility to publish, a high standard of work presented is also indispensable. In general, the elimination of political restrictions has not resulted in the emergence of a new group of promising authors. The mechanisms of international scientific collaboration are largely oriented towards well-known scientists whose contributions to science are already assured.

5. Involvement in International Co-operation and the National Achievements of Scientists

The results of the Russian survey study permit us to go beyond statements referring globally to science in a given institute, or a category of institutes or a country, and to add a micro-level view to the foregoing remarks about the changing extent, geography and forms of international co-operation. The survey sample which deliberately sought a high concentration of active participants in international interactions made it possible to distinguish five groups of scientists, labelled A to E, according to the degree of their involvement in international interactions. The indicators used were: publications in foreign journals etc., papers presented at international meetings, involvement in joint projects, grants awarded by Western foundations and agencies, and use of high speed electronic communications.

Group A scientists are internationally recognised, publish regularly abroad and present papers at international meetings; they obtain Western grants or participate in joint projects with Western partners, often as leaders of research teams. Group B scientists have not yet attained full international recognition. They participate in Western grant-based or joint projects and are continuously searching for new opportunities. However, they rarely publish abroad, and their participation in international meetings is haphazard, even though they make considerable efforts towards this end. Group C scientists have no publications abroad and very rarely attend international meetings, but they are involved in international interactions as members of a joint project team, having no grants of their own. Scientists in Group D report a very weak involvement in international interactions; their foreign contacts are random episodes which do not significantly affect their professional lives. In Group E, finally, there is a complete absence of international contacts. The five groups were subsequently analysed with respect to a large number of personal characteristics, professional activities and attitudes (using over thirty cross-correlations). The hypothesis was that there would be a positive correlation between involvement in international collaboration and the general research efficiency of scientists.

Though the non-representative nature of the sample does not permit us to draw quantitative conclusions about the relative size of the five groups among academy scientists in Russia (let alone in other CEE countries), the survey results provide interesting findings. Far from being mere statistical categories, the five groups each have a very specific character and can, in fact, be considered to represent different strata in the academic community.

Group A scientists are a genuine elite.[19] They not only lead in international interactions, but they are also highly efficient in the national science system (by way of publications, grants received, etc.). Critical and realistically-minded in all their assessments, they report the highest involvement in additional work in science, and are the group least engaged in non-scientific, additional activities (8%). Their research receives better technical support; they take advantage of the best opportunities for travelling abroad; and they remain in science because they see a real opportunity to pursue research and are unwilling to change their mode of life. While being more painfully aware of the disruption of former collegial links, they nevertheless more often claim satisfaction with their work (33%). Nearly half of the Group A members are research associates; heads of research sub-divisions are only slightly over-represented, while top managers are practically absent.

Group B scientists, initially deemed to be tomorrow's leaders, the future elite, may in fact be only "specialists in international collaboration". They are quite active in international interactions and in some respects even the group most favoured by Western partners (e.g., with respect to stays for training purposes and computer supply). But the scientific activity of Group B members in their home country does not meet expectations – publications are limited to articles, their additional engagement in science is low, while their non-scientific engagements are at a maximum (28%). Though their access to technical facilities is practically the same as for Group A, their estimates of their prospects to maintain the quality of their research was the lowest of all the groups (4%). It seems that the majority of Group B scientists have not sufficient potential for long-term successful work, and their involvement in international interactions is situational rather than essential.

Group C scientists are not leaders, but self-confident professional performers. No research would be possible without them. They are active and efficient in various respects, reasonably critical, and optimistic. Reporting a poorer provision of necessary facilities than Group B, they yet feel more comfortable and stable than all other groups. Their concentration on science and their level of work satisfaction is practically the same as in Group A , and their confidence in the opportunity to pursue their work is even greater.

Scientists of Groups D and E are similar in many respects. Their weak involvement in international interactions (if any) is not compensated for by their efficiency in national science; objective evidence of their rather low professional success surprisingly combines with their very positive estimates of the state of research in their own sub-divisions and a very high rating of the quality

19 The characteristics of the group as a whole do not necessarily hold for all its members.

of their own work. However, there are some important attitudinal differences between groups D and E.

Group D members stress the existence of good opportunities (satisfactory experimental facilities, sufficient opportunities to pursue research and travel abroad, etc.). At the same time they manifest the maximum level of dissatisfaction (67%). There is evidence of a really high professional potential in at least part of the group. Group E, in which sub-division heads are strongly represented, has the lowest index value of professional efficiency, and the lowest estimates of future prospects. Nevertheless, group E members show the minimum dissatisfaction with their own research activities (52%).

Contrary to expectations, analysis of the groups shows that no clear relationship exists between the involvement of scientists in international interactions and their professional performance. An active involvement in international interactions is not a conclusive indicator of a researcher's professional efficiency (see Groups B and D). Group A, Group C and Group E scientists actually preserve their previous patterns of professional behaviour, only modifying them to adapt to the new conditions of *open science*. Group B reports an evident predominance of international activity. However, being to a certain extent situational and in part even simulated, this activity cannot be fruitful in the long run. Productive scientists of Group D, in contrast, lack foreign financial support, while part of the means granted to Group B are spent inefficiently. Group D scientists are largely active within national science and link their professional ambitions only with the latter. Thus, international interactions and professional performance go together only in Groups A and C.

The survey findings also point to changes in the stratification of the scientific community. The differences in the composition of the elite groups (for example, heads of subdivisions were underrepresented in Group A, while the proportion of research workers was large) suggest that a new type of stratification is evolving within the professional community.[20] This new stratification did not, however, emerge on a blank sheet. It reflects the "genuine" stratification of the scientific community that has always latently existed, based on professional competence rather than official standing. The new stratification starts from the old one: each former stratum is split, depending on the support that scientists manage to get; scientists who achieve support preserve their status (or sometimes even raise it), while those who fail lose status. As a consequence, the dead wood which has always constituted a considerable part of the Soviet scientific

20 Previously a scientist's rank in the Soviet stratification of science was largely determined by his/her *official position*. Advantages offered by the official position were not to be questioned, while the barriers between official positions were insurmountable: a scientific worker could never obtain a reward which the head of his/her subdivision, or even institute, had not already received.

manpower gradually becomes isolated, sinking to lower levels and losing its future prospects.

6. Concluding Remarks

The issue of international collaboration inevitably has a political aspect, being related both to current economic considerations and to the continuing opposition between *patriots* and *Westernisers* that is particularly apparent in Russia, but traces of which can be found in all CEE countries.

At present it is hard to tell whether science is undergoing a fruitful transformation in the CEE countries or whether it is in the process of being destroyed. Against this background, different opinions exist about the role played by international collaboration. Those viewing the current situation as a process of decay of the national science system either believe that foreign support slows down its erosion and enhances the chances for revival, or they believe – quite in contrast – that it undermines the potential of the national scientific community by inducing a brain drain and by cultivating Western attitudes and alien patterns of research organisation. The optimistic view of the changes currently underway is usually accompanied by the opinion that international collaboration has a positive impact.

The form of interaction with international science that attracted the greatest attention throughout the earlier part of the 1990s was direct involvement in research in advanced countries, i. e. long-term visits to participate in joint projects (at the initiative of the host country), contract-based work abroad, or emigration. Accordingly, the phenomenon of the brain drain is widely discussed in most CEE countries. The case studies did not approach this subject explicitly, although the Russian survey study did. While the subject does attract a great deal of attention in Russia, the survey results showed an evident lack of intention to leave the country and a clear orientation toward collective research. This invalidates the widespread fear that international interactions promote emigration and destroy the collective spirit of Russian science. In the elite institutes of the Russian Academy of Science, only 2% of the respondents stated their intention to emigrate in answer to a corresponding question – which at the same time provoked immediate additional remarks that such opportunities do not exist. In recent years, the interest in emigration has obviously been superseded by a shift to "pendulum" migration. In part this may be a function of the scarcity of real opportunities, which may well be better for Polish or Hungarian scientists, who, in contrast to their Russian colleagues, had already been able to maintain close relations to Western science prior to the political transformation. Another factor that plays a role here is the Soviet tradition of collective work. Many Russian scientists feel that they waste their professional potential if they are away from their native teams, and this prompts deep dissatisfaction.

With respect to the potential loss of the collective spirit, it is interesting to note that practically all (93%) of the male scientists covered by the Russian survey expressed their willingness to participate in collective research; the interest in receiving individual grants reported by some of them is not seen as an alternative, but as supplementary. True, female scientists more frequently expressed a negative attitude toward collective research and an orientation exclusively toward individual grants (18%), but this reflects their dissatisfaction with the subordinate role often assigned to them in collective work rather than an essential disappointment with collective research *per se*. However, the tradition of collective work will not be able to withstand the atomisation of research referred to before. This is a process that is taking place in all CEE countries.

The extraordinary development of international interaction in science has made this a subject of wide interest and much speculation. The long-term effects of the developments described in this chapter are not entirely clear. Doubtless, however, international collaboration plays an essential role in the current functioning, and in the transformation, of science in post-communist countries.

References

Science & Engineering Indicators – 1987. Washington: NSB, 286–87.

International Science Foundation (1994). *Annual Report*. Washington: ISF, 1994.

Nauchno-organizatsionnaya deyatelnost'v 1994 *g*. Godovoj otchet Presidiuma RAN. M.: RAN, 1995, 93.

Mirskaya, Elena (1994). Scientists: About Their Present and Future. *Herald of the RAS*, Vol. 64, Nr. 5, 341-348.

Mirskaya, Elena (1995a). *Die Rolle des Internationalen Zusammenwirkens in der Transformation der Wissenschaft Russlands* (research report for the Network workshop, Berlin, December 11-12).

Mirskaya, Elena (1995b). Russian Academic Science Today: Its Societal Standing and the Situation within the Scientific Community. *Social Studies of Science*, SAGE, Vol. 25, Nr. 4, 705-725.

THE TWO-EDGED SWORD OF AUTONOMY: CHANGES IN THE ACADEMY-INSTITUTE RELATIONS

Kostadinka Simeonova

Centre for Science of Science
Bulgarian Academy of Sciences
Sofia – 100

1. Introduction

Reforms at the various Academies of Sciences in the East and Central European countries are each proceeding at their own unique pace. All of them, however, are in the process of dismantling the strong centralisation they inherited from their various totalitarian systems. In the past, academies exercised minute bureaucratic control over the detailed activities of their incorporated institutes. Centralised planning, financing, and inspecting used to be established practice for managing research activities. Furthermore, this kind of centralised control also reached out to include strict regulation of partnership and contract activities, personnel policies, the regulation of researchers' careers, and avenues for international collaboration and information exchange. As a rule, institutes have been excluded from decision-making processes within the academy.

The most essential change to take place in managing scientific research after the start of reforms has been the acceptance of the principle of autonomy in science. In some countries, universities and their academy have enshrined this principle through legislation. In these cases, the charters of the academies have been changed and internal structural reforms have stemmed from that. In other countries, although legislative change is delayed, similar reforms are under way in the academies within the ethos of more general political and economic change: the process of strengthening the autonomy of the academies' structural units is ongoing. A new type of relationship between the academy and its institutes is emerging on the basis of moves towards decentralisation and of a reduction in hierarchical levels of decision-making. Thus, institutes of those academies have, in practice, more room for manoeuvre.

It must be pointed out that these changes take place in the context of a survival strategy which is accepted both by the academies and by their institutes. This affects their mutual relations in interesting ways. In some cases, autonomy is being imposed upon institutes rather than being a response to a felt demand.

R. Mayntz et al. (eds.), East European Academies in Transition, 125–139.

In these instances, the virtual imposition of independence becomes a tacit means for the academy to enact its survival policy.

As already mentioned, there are differences in the pace of reforms between the several East European countries. In all except Poland, institutes have legal autonomy and this has undoubtedly increased their independence. Legal independence has had the most dramatic impact in countries where the restructuring of the academies has taken the most radical forms, viz. the Baltic states of Lithuania, Latvia, and Estonia. The academies in these countries have shifted to the West European model of "learned societies". As such, developments in the Baltic states are being closely watched by fellow East European academies since they constitute a "natural experiment" in guaging the effects of radical – as opposed to evolutionary – transformation. Poland's route towards reform also provides an interesting case for observation. The Polish Academy has been entirely stripped of its financing functions with respect to its institutes, and these organisations have now been placed under the dual authority of the Academy and the Committee for Scientific Research.

Our study provides an excellent opportunity to review and assess the initial results of these various attempts at managing reform. The use of documentary sources together with experts interviews enables us to shed light on changes in academy-institute relations and to offer some explanatory suggestions on our findings.

In this chapter we address the following questions: What general organisational changes are emerging from the new relations between the academies and their institutes? Which of the former functions of the central bodies of the academy are preserved and which abolished? In the process of establishing new relations, what are the main sources of tension to emerge? In the conclusion, we discuss prospects for the further development of academy-institute relations.

2. Organisational Changes and New Forms of Interconnection Between Institutes and the Central Bodies of Their Academies

Establishing Institutes' Representation Within the Management Structure of the Academy: Two main forms of representation were found. One form is the representative role played by the Committees (or Councils) of Directors of the institutes, which aim to improve the co-ordination of the institutes' activities. The other form is efforts to increase the direct participation of reseachers in the supreme body of their academy – its General Assembly – in handling decisions over research policy, regulations, financing etc. Although representative bodies sometimes carry the same name, their functions may differ in respect of their formal and influential capacity. For instance, the Council of Research Institutes in Hungary is attached to the General Scientific Secretary of the Academy and its main function is to co-ordinate future activities of the institutes. In Poland, by contrast, the Council of Research Institutes is an informal consultative body from which the institutes' administration (i.e., its directors and scientific secretaries) is excluded, and it represents the interests of the reseachers independently from the Secretariat of the Polish Academy and its subordinate structures.

Obviously, in the period of transformation, more weight is given to official representatives of an institute (who are usually elected by that institute's members) in the supreme body of their academy, as is the case in Belarus and Bulgaria. In Bulgaria, the General Assembly is entirely peopled by its institutes, who each have a specific quota of representatives. This Assembly has exclusive power to negotiate structural change, to allocate resources, and to participate in a whole range of policy issues. In contrast to Belarus, it is separate from the Assembly of the Members of the Academy, i.e., academics and related staff.

Setting up Mechanisms for Evaluating Institutes: Following public criticism of over-staffing, the initial steps of the academies in almost every country under study were directed towards sharp staff reduction within their institutes. Towards this end, evaluation procedures of differing type and scope were implemented. In some countries, evaluation has led to the ranking of institutes such that they become eligible for graded levels of funding (Poland, Slovak Republic, and the Czech Republic). In Bulgaria, evaluation has been mainly used to effect structural changes irrespective of the financing of the separate institutes. In some countries, *ad hoc* commissions have been created although specific evaluation procedures are regulated by the academies' charters. In other countries (the Czech Republic and Slovak Republic), the Academic Appreciation Committee is a permanent body established outside the academy which includes foreign members (40%). Another peculiarity is that in some countries, researchers and the institutes which employ them are evaluated as a single entity (the Czech Republic, Hungary, and Slovak Republic), whereas in others, evaluation occurs at two levels, i.e., a general assessment of the institute's activity and a subsequent assessment of individual researchers within that institute (Bulgaria).

Relinquishing Control Over the Organisation of Research Programmes: In the past, an important means for academies to influence activities has been for them to control specific research programmes. The central bodies of the academies provided their respective institutes with an organisational and financial framework to support pre-chosen avenues of research, e.g., priority problems in economic, environmental, and regional affairs. However, at present, only the Romanian Academy and the Czech Academy – through its Programme for the Development of Fundamental Studies which was set up in 1995 – pursue overarching research programmes. In the former case, the relevant specialties of the institutes are included, and participation in the programme amounts, in effect, to an indirect evaluation of that institute. In the remaining countries under consideration, this practice no longer occurs.

3. Main Changes in the Academies' Functions With Respect to Institutes

Strengthening the Autonomy of Institutes: The experts, leading researchers, and heads of the institutes who took part in the study express strong satisfaction with the main thrust of the changes: the strengthening of the autonomy of the institutes as part of the academies' transformation.

Weakening of Bureaucratic Control by the Central Bodies of the Academy: This general trend is manifested in two main ways: reducing the paperwork adressed to the academy, and the academy's relinquishing of binding control over both the planning and detailed execution of institutes' activities. In practice, the yearly plans – formerly, a management tool for controlling the distribution of monies – have been abolished in all the academies. Presently, annual reports are rather of the character of systematic accounts of the institute's activities, whose main aim is to legitimise it before the scientific community and the public at large.

Institutes are entirely free to change their research topics without submitting them to the approval of the planning bodies though, again, we found our interviewees' experience of this general change differing from country to country. Thus, respondents from the Ukrainian institute of physics point out that currently there are less barriers between institutes and more freedom to choose partners within the academy whenever joint projects are being negotiated; formal relations have been weakened at the same time as informal networks have been strengthened. In other countries (Hungary, Lithuania, and Bulgaria), respondents remark that the thematic profile of their institute has never been dependent on the academy and, in consequence, they are not undergoing any substantial change with respect to this reform.

The General Statutes of the Academy Diminish in Importance Relative to the Internal House-Rules of the Institutes: Poland offers us a good case in point. Since 1970, the statutes of the Polish Academy which prescribe detailed procedures for the functioning of its institutes (e.g., the election of directors and of scientific boards, etc) have remained unchanged. However, Polish institutes tend to disregard these edicts and have been acting in accordance with new house rules of their own, bringing them into sharp conflict with the Academy's statutes. In academies where new statutes have been written since 1990, principles such as decentralisation and the autonomy of institutes to shape their own internal regulation have been explicitly written in. In the past, bringing individual institutes' rules into conformity with the general statutes of the academy was accomplished by the so-called "type codes" for the institutes. Using these type codes, the old centralised model of academy-institute relations was compulsorily reproduced. This practice no longer occurs.

Interviewees seem satisfied by these changes on the whole. Our respondents endorse the greater freedom of their institutes to make decisions – albeit that this process has carried some organisational inconvenience in its wake. For example, according to a Lithuanian expert, "the autonomy of the institutes did not substantially affect the freedom of decision-making in the institute, but did contribute to the increase of its organisational troubles" (Lithuanian institute of semi-conductor physics), since, after their separation from the academy, the institutes are now charged with responsibility for dealing with issues new to them such as financing, buildings, supplies, transportation, etc.

However, we also encountered some negative assessments of the change process. According to some respondents, it raises questions about the precise nature of the academy's duties towards their institutes, and some confusion remains as to what those institutes are entitled to expect from their academy. Thus, a Russian interviewee claimed that the academy "has completely abandoned its institutes" (institute of biology), and that it now simply operates to preserve its central bodies without taking sufficient care over the future development of its research units. Similar criticisms are made in others of the institutes under study in Belarus and in the Ukraine, where respondents consider their respective academies to be insufficiently involved in the development of science policy and strategic goals. In Bulgaria, our interviewees queried the priority-setting of the academy, most especially "the lack of balance in the policy towards the different kinds of research carried out at the institutes" (nuclear research).

Academies Lose Their Significance as the Main Funding Source for Institutes' Research Activities: In several academies, the main reason for a general weakening of control stems largely from this particular change. Increasingly, academies are not able to provide institutes with research resources either for infrastructural development or for more project-specific running costs. This diminishment has followed a specific history in each country. In Poland, following the establishment of a special Committee for Scientific Research, institutes of the Polish Academy are financed directly by this organisation. In contrast, in most of the other countries under study (Russia, Hungary, Bulgaria, and Romania), alternative financing institutions have emerged which provide some national funds for scientific research, together with local specialist arrangements for the financing of any given project. Whereas in the past academies had the exclusive right to dispose of funds, nowadays that role has decreased to the point where its main financial responsibility ends with the provision of salaries for researchers and specialists in the institutes.

Our study offers evidence for how academy-institute relations have been altered by the changes in the management of funds. Thus, in the Academy of Belarus, once 25% of the available financial resources were placed in the hands of its departments, the significance of those departments to the Belarussian institutes rose dramatically, while the role of the Academy's Presidium commensurately decreased. In academies where grant agencies have been created from within (the Czech Republic, Slovak Republic), they remain important players precisely because of their continued financial importance to their institutes.

In this situation, the direction of an institute's research activity, the renewal of its infrastructure, and the kinds of relationships which can be forged with other partners, depend exclusively on the institute's own initiatives. Locally-developed survival strategies turn out, in practice, to be much more effective than the over-arching strategy of the academy which is confined, in most cases, to the distribution of a modest state subsidy. As a consequence, institutes are more at liberty both to pursue research funds through all sorts of local arrangements, and to direct their own international participation in projects.

Clearly, institutes differ amongst themselves in this respect. As regards the speed of change, we found much greater heterogeneity amongst institutes, than in the academies taken as a whole.

3.1 FUNCTIONS RETAINED BY THE ACADEMY WITH RESPECT TO ITS INSTITUTES

The attitude taken to the academy in Eastern and Central Europe is determined to a great extent by its place in the national science system. We refer here not only to its past privileged position, but also to the ongoing processes of transition in science, namely, to the decline in industrial science and to the considerable weakening of university scientific activity. Therefore, our respondents were unanimous in stressing that the academy – despite its undeniable shortcomings and slow pace of reforms – still remains the mainstay of research activity for its institutes.

Many of our interviewees expressed apprehension that institutes which leave the academy will find themselves out in the cold financially, however weak or insufficient they regard the academy's financial aid to be. This fear was expressed by respondents in Hungary and Belarus. One pointed out that "the organism of the Academy of Sciences is preserved due to the existing organisational structure. If the Presidium is eliminated, the Academy will break down into islands. Then its future existence will be vague. It seems better to preserve these structures as a fortress which helps us to survive" (Belarussian institute of physics).

As mentioned earlier, some countries endorsed radical change, i.e., they transformed their academies into a Western style, based on societies of outstanding scientists, and their respective institutes acquired the status of independent state research organisations. According to our study, these countries' practices justify the fears expressed above. In Latvia, where institutes gained independence in 1992, our respondents judged their splitting from the academy to be "premature": in one institute, an interviewee pointed out that links to the academy – when they are cemented only through some individual researchers also acting as members of the academy – are insufficient. In a second institute, it was noted that the interests of the institute with respect to relevant bodies were better represented by the academy, than after reforms took place. Scientists from the institutes we studied considered their new position – which locks them into a closer dependence on the Ministry of Science – as more unfavourable than pertained in the previous regime. This reassessment corresponds with a more recent tendency to restore part of the academy's functions with respect to its institutes. As a result, new agreements between the academy and its institutes were signed in 1995.

A frame agreement of the same type is also employed by the institutes in Estonia. This allows for a so-called "association" which permits an institute to enter into new relations with the academy, which carries certain rights and duties. Reading the detail of these agreements, which are entered into voluntarily by the institutes, we can draw certain conclusions about the real role of the academy in the institutes' activities. Estonia, in 1995, has taken the same path as Lithuania and Latvia before it: it has abolished the Soviet type of academy, and has replaced that with institutes which have the status of independent state organisations. The current relations of academy and institutes are based on bilateral agreements which articulate the duties of the academy. The duties of the Estonian Academy, stated below, represent its role with respect to the newly-independent institutes. According to the agreement, the Estonian Academy is obliged:

- to inform the institutes about its decisions and international collaborations;
- to represent the institutes' interests'
- to defend the institutes' interests;
- to account for the institutes' achievements in the Academy's annual reports; and,
- to include members of the institutes in the Academy's activities.

Estonia offers one example of a realignment of academy-institute relations, subsequent to the first major shift towards institutional autonomy. However, our study shows a similar pattern being enacted across a number of Eastern and Central European countries. We may summarise the functions which the academies have retained, or regained, as follows:

The Strategic Function: The academy retains the role of a focus for the pursuit of basic research in its country. For decades, the academy encapsulated all the relevant expertise needed to pursue that research, and it was the academy which carried social responsibility for enabling its continuance. In the East European countries, after the start of the reforms, sharp debates arose as to the fate of such fundamental research. In line with Western European practice, the opinion has been expressed that, in small countries, basic research should be conducted by the universities. In this context, it is understandable that institutes of the academy, in their struggle for survival, should turn to the academy which has traditionally had a leading role in guiding basic research. Most of the institutes covered by the study expect their academy to energetically elaborate and realise policies which enable basic research to be pursued.

Part of our respondents' reservations about the current *status quo* is that the central bodies of the academy have not been active enough in this respect. The point has been expressed that the market context, when combined with administrative control being placed in the hands of individual institutes, cannot replace the academy as a driver for basic research.

In some countries, the successful accomplishment of this function is supported by grant agencies established within the academy (the Czech Republic, Slovak Republic). Here, not only are priorities set by the academy but, in addition, it formulates a strategy for particular fields of study. Interviewees from Slovak Republic made an interesting observation: they remark that the academy's autonomy is on the decline although, unusually, there has also been a budget increase for 1995. Thus, the academy's policy in Slovak Republic appears to be sufficiently well-funded to ensure its success.

The Process Function: This function concerns the management of research, and the means of communication and collaboration, in the production of scientific knowledge. In this respect, the academy is well-placed to organise interdisciplinary and multidisciplinary research. Despite the dangers of newly-autonomous institutes becoming isolated from each other – which was mentioned by respondents in Romania and Belarus – our interviewees thought that the academies offer good opportunities for facilitating collaboration between institutes.

In some countries, this collaboration is largely accomplished within the respective divisions of the academy (Poland, Hungary) since they include units from closely-related research disciplines. Not only do these divisions organise discussions on common topics, they also take responsibility for the quality of research in their institutes by means of periodic assessments and work reviews. In some cases, they control the distribution of a significant portion of the academy's budget (25% in Belarus). In Bulgaria, institutes share the same buildings which encourages some inter-institutional collaboration on a range of practical issues. Although interdisciplinary research programmes are less common than before, it remains true that, in some countries, the academy's control of general funds remains useful in supporting such projects. Therefore, in most of our cases, interviewees stress the continued importance of collaboration between institutes and their academy.

Particularly important – and repeatedly stressed by all participants in the study – are the international activities of the academy. By numerous bilateral and multilateral agreements, the academy generates a favourable environment for its institutes to find international partners. It should not be forgotten that in many cases the academy arranges favourable conditions for collaboration with fellow academies and other international organisations, e.g., through funding on a "no-cost" basis. In the period of transition as a whole, the international scientific activity of the Central and East European countries has been strengthened. In this respect, academies have become national information centres which facilitate the participation of institutes in the programmes PHARE, PECO-COPERNICUS, TEMPUS, etc. An interviewee in the Ukrainian Academy pointed out that its departments are turning into *sui generis* information centres on international funds, projects, joint studies, etc. This role has been strengthened at the expense of their previous preoccupation with the administration and management of their institutes.

The Protective Function: The academy represents and protects the common interests of the institutes when negotiating with state bodies about proposed legislative change. In that kind of instance, individual institutes cannot defend their interests as efficiently as the academy can. Academies enjoy full participation in all the representative state bodies, through the advisory bodies which inform the legislative and executive tiers of government. In turn, the state seeks out the academy's expert opinion to guide scientific work. For instance, the General Assembly of the Bulgarian Academy closely participates in all decisions dealing with national science policy, altering such policies in the light of its institutes' research strengths.

The Legitimising Function: The academy possesses a rich cultural repertoire to legitimise the activity of its scientists. It is associated with traditional values such as academic liberty and the critical function of science, together with diversity of views and the free development of alternatives which are important preconditions for the creation of pluralist societies. It is widely acknowledged that even in an anti-science cultural milieu, the historic role of the academy aids the process of national revival. For many years, but especially during the demise of the old regime when economic turmoil was acutely felt, the academies still managed to legitimise the activity of their institutes by means of highlighting their practical usefulness.

In the current context, our study shows how there is a shift towards basic research, legitimated by drawing on values such as the necessity to participate in international scientific trends. Our interviewees do not always approve this change but, at the same time, they take into account the confusion which results when a society does not know what its science is "for"; when there is still only a limited public demand for immediate practical science. As a whole, it is beyond dispute that the academy has great potential to counter unfavourable social attitudes, thereby creating greater stability for the institutes to go about their work.

In this respect, the structure of the academy in which practically all research endeavour is represented – natural sciences, technical sciences, social sciences, and humanities – enables it to launch new ideas, and to defend its practical objectives and achievements, as well as its conceptual contribution to the processes of reform in society.

The Symbolic Function: This function relates to the ability of an institution to present itself as an effective force with respect to its internal or external environment. Belonging to the Academy of Sciences symbolises the quality of an institute's research. As a result of previous historical developments, the academies have become national reservoirs of highly-qualified scientific talent, and they have established themselves as leaders in forging international partnerships. Elite departments, generously subsidised by the state, belonged as a rule to the academy. Despite the pressure to keep up the quantity of material produced, these departments have preserved high performance standards and achieved high publication rates in leading international scientific journals. Thus, possessing the mere title of belonging to an institute of the academy feeds into the "credibility cycle" and favours the image of the institute. This, in its turn, has a considerable knock-on effect in establishing scientific contacts at home and abroad, and in favouring that department's ability to win grants and receive state support.

This dimension of academy-institute relations has been convincingly confirmed by our study. Institutes which have legally become independent state research organisations still preserve, in their titles, the attribute "Institute of the Academy". This is the case in two of three institutes studied in Latvia (institutes of physics, polymer mechanics, and organic synthesis). The democratic decision of all scientists in one institute was to retain the name "Institute of the Latvian Academy of Sciences". The main argument in favour was that the academy is thus preserved as a protective "umbrella" for the institute, particularly with regard to how it is assessed in the future. In our interviews, scientists point to the fact that being evaluated by the academy ensures that performance is judged on scientific criteria, and in a far more objective way than would be the case if the adjudicating body were the relevant Ministry.

At the same time it has to be stressed that the symbolic function of the academy depends to a great extent on how its credibility is perceived by its own institutes. Referring again to the case of Latvia, it is interesting that one of the studied institutes no longer claims any association with the academy. This can perhaps be explained by the fact that historically this elite scientific institute was more strongly championed by the Academy of Sciences of the former Soviet Union than by its own native Academy. The former still carries much more weight in the eyes of that Latvian institute than the Latvian Academy does. For this reason, the institute prefers to keep its distance from the Latvian Academy, notwithstanding the massive political changes which have occurred in the former USSR.

An opposite example is offered by the case of Romania: the closing of the academy in the '70s has lent it the air of being a victim of the former regime. Therefore, identification with the Romanian Academy is useful for its institutes in the present political context. Our Romanian respondents showed deep reluctance to be identified with the conservative and bureaucratic Academy of Russia, and consequently endorsed the most radical form of transformation, i.e., the separation of institutes from the academy and a remodelling along Western lines. As already mentioned, the pace of reforms in the Academy of Russia is slow, and this may account for the strong criticism levelled at that particular academy. On the other hand, some Russian respondents pointed out that their criticism is mainly directed at particular individuals in the academy: the policies of the Academy of Russia are a source of dissatisfaction because they have a personal dimension.

4. The Academies' Tacit Means of Influencing Institutes' Activity

According to our respondents, each academy's influence on its institutes has changed from a direct to an indirect one, based primarily on financial regulation rather than on direct interference in its institutes' affairs. The most significant forms of influence are considered here.

Evaluation by the Academy as the Basis for Restructuring Institutes and for Allocating Funding: A range of academies evaluated their institutes at the beginning of reforms. In Hungary, the Czech Republic, Slovak Republic, Poland, and Bulgaria this was done by means of establishing special bodies or by *ad hoc* commissions, and all units have been subject to review. In some countries, this has been achieved with the participation of international experts, e.g., in the Hungarian Academy of Sciences. The results of an evaluation have direct implications for the survival or closure of a unit (Bulgaria, the Czech Republic); for how units are categorised and, therefore, for how they fare in slicing up the budget subsidy (Poland, Slovak Republic); and for the reduction of the institute's personnel. In the present context of stronger dependence on budgetary income, institutes are rather sensitive about the evaluation and its consequences. In certain cases (mainly in the social sciences), interviewees expressed doubt about the objectivity of the evaluation criteria and were of the opinion that their institute had received biased treatment on ideological grounds (Poland, economics). This state of dependence is likely to deepen in the future since, according to the new statutes of the academies, evaluation has become a permanent mechanism for research management.

Determining the Institutes' Activities: Reforms have brought considerable autonomy to the institutes in determining their choice of research topics. Nevertheless, there are barriers to how far institutes may diversify their activities because, as the new agenda evolves, it is becoming clear that the academy wishes to preserve certain activities and to curtail others. When the academy became the guardian of property which had previously belonged to the state, the academy, in the first instance, gave over the use and management of that property (i.e., the renting of buildings, devices, terrain, etc.) to its institutes. After 1993, however, we note a change in the academy's attitude. In particular, we find a concern about the growing commercialisation of research results and the setting-up of spin-off companies which exploited the resources and the property of the academy. Across a number of countries, the academies took stiff administrative measures to expel such nascent firms.

Our study shows that dissatisfaction with these measures varies from country to country; we found quite a calm reaction from respondents in the Hungarian Academy, in contrast to a sharper one in Bulgaria. According to one Bulgarian respondent (institute of molecular biology), the expulsion of a firm from the institute has been "a great mistake, because at the moment the firm in question is doing very well, due to some foreign investments. Since the institute has sold its share, its losses are twofold: in the form of intellectual property and of potential income".

The academies' moves against commercialisation reflect more than a mere relapse into an outdated bureaucratic mind-set. Commercialisation of research findings portends the erosion of classic scientific norms, and utterly alters the traditional shape of the academy. More pragmatically, given the uncertain flux which accompanies economic change, commercialisation permits the ministry of finance to apply additional restrictive measures on the academy and to further reduce its state subsidy. Therefore, there are reasons to contend that the academies' restriction of their institutes' commercial activities simply reflects part of their own survival strategy – a strategy which protects the common interests of most of its institutes. The same rationale may explain the remarkable prevalence of fundamental research which we found in the period under study. In our opinion, the academies' preference for research activity over commercialisation (especially in the technical sciences) represents a means for those academies to retain tacit control over their institutes. In the present context, institutes seem willing to continue to structure their activities according to the academies' wishes, so that they may remain incorporated.

5. Tensions in Academy-Institute Relations

Trends towards institutional autonomy are necessarily accompanied by certain tensions and unresolved problems. In this section, we point out the important issues which emerged from our study.

There is a gap between the declared aim of promoting autonomy, and realising that aim in practice. This was pointed out by respondents in Hungary and in Bulgaria, and we also noted some instances of a return to more centralised practices, e.g., in the handling of staff vacancies, and of distributing monies to support international collaboration.

There are discrepancies between the formal regulation of academy-institute relations, and their operation in practice. For example, in the Baltic countries where institutes have the formal status of independent state organisations, the final decision about membership on the management bodies (i.e., the Director, and the Scientific Board) is nevertheless taken by the academy. (In Estonia, elected members must be approved by the relevant department.) Paradoxically, therefore, the most radical model of renegotiated academy-institute relations provides the Baltic institutes with less autonomy than the more evolutionary varieties which we found elsewhere in East and Central Europe. For example, in Bulgaria, an institute's Scientific Council is elected by the General Assembly of that institute's scientists, and their choice is not subject to veto by the academy.

Delays in enacting legislative change may cause problems. In Poland, institutes do not have any legal status yet their internal house rules have evolved to a point where they contradict the Polish Academy's charter. One interviewee (institute of social sciences) pointed out that although the Academy's Secretariat is sympathetic to how institutes frequently exceed their formal powers of decision-making, nevertheless such sympathy is necessarily subjective. The lack of formal, legal, regulation of their relationship leaves everyone in limbo.

In certain cases, formal democratic procedures are a facade which only pay lip-service to change: in practice, the old relationship continues. One respondent remarked that "only those who will obey get into elected bodies. Formally, this is undoubtedly a democratic step. However, the only ones elected are those who will act under the dictates coming from above". (Belarussian institute of biology).

Occasionally, there are serious, pragmatic, constraints on altering academy-institute relations. One of our interviewees in the Estonian institute of chemistry pointed out that the institute remains the academy's property – lock, stock and barrel. Were they to separate from the academy, they would lose everything.

There are queries about the criteria used for evaluating institutes' performance. While evaluation *per se* is viewed positively, some problems with it have emerged. Institutes in Romania and Bulgaria are worried that their academies' evaluation process does not take sufficient account of their research specialisms, nor of the general thrust of science policy in their institutes in the period prior to the evaluation. This conviction was especially marked in the technical sciences institute attached to the Bulgarian Academy. We already noted that the Polish institute of social sciences feared that their evaluation was biased, on ideological grounds. Apprehensions of this kind hinder the development of a relationship of trust in academy-institute relations, and diminish the value of the evaluation procedure.

Some of our respondents have contradictory feelings about the general move towards autonomy. Certain interviewees in Russia and the Ukraine still echo the paternalist ethos of the past, and blame the academies for not "taking sufficient care" of their units. It is clear that the state's exit from the academy system has been a profound shock which has stimulated two kinds of reaction: one towards striving for survival through fragmentation and autonomy, and the other towards searching for a collective strategy to meet the challenges of transition. Our study indicates that the emphasis of response to change differs across countries and institutes. On the one hand, strong institutes are able to meet the challenge of change on their own: a respondent in an elite institute in Bulgaria (institute of molecular biology) considers that only weak institutes suffer from reforms which support moves towards autonomy. On the other hand, an interviewee in another institute (Romanian institute of biochemistry), argued that autonomy is over-rated: "Our autonomy consists in allowing us to work without finance. Perhaps it would be better to have less freedom. The Academy pays us our salaries, approves our plans, and asks what we have done at the end of the year. This is too little".

6. Conclusion

Academy-institute relations are undergoing profound change in this transition period. The process of transformation is not complete, and is proceeding at different rates in different countries. Nevertheless, our study identified some common trends which we now summarise.

The most important change has been efforts to promote the autonomy of institutes. This is being accomplished by decentralising decision-making powers, and by placing a higher proportion of available resources at the personal disposal of the institutes.

Institutes' autonomy is not inconsistent with the academies' drive towards survival. Autonomy offers the whole system greater flexibility in adapting to economic uncertainty, although this demands excellent management skills.

In the changed context, academies are relinquishing a good deal of their control over the financing of research, although they continue to meet the costs of salaries and of infrastructural maintenance. This is having the effect of weakening institutes' dependence on their academies, and of decreasing the academies' administrative functions.

Academies still retain their legitimising function with respect to basic research. They continue to act as an umbrella for their institutes such that they are in the best position to protect institutes against political, social, and marketplace uncertainties.

Institutes have high expectations, and make real use, of the academies' established international contacts. The academies continue to embrace their representative role in a range of international scientific programmes and organisations.

In the whole, institutes are positive in their assessment of the academy system. They value its role in enabling fundamental research to thrive. Indeed, our respondents often stressed how difficult it would be for an institute to survive successfully, were it to detach itself completely from the academy system.

The formal regulation of the changes just noted frequently lags behind the reality, since the changes are occuring from the bottom-up. Academies are more frequently driven to accept rather than initiate emergent shifts in academy-institute relations. New fora are being created to facilitate dialogue and, almost everywhere, these embody the principles of meritocracy and representation. In this way, institutes are coming to participate in decision-making and in management of the academy system itself.

Because of the legacies from the past and the necessary uncertainties which attend transition, change is accompanied by tensions and difficulties which differ in their detail across countries. However, the overall direction of change is irreversible. The institutes are aware of the advantages of remodelling out-dated academy-institute relations, and they see their task as improving the academy system as a whole.

SELF-CONCEPTION, STRATEGIES OF TRANSFORMATION, AND MOODS IN COPING

PETER WEINGART

Faculty of Sociology
University of Bielefeld
D-33615 Bielefeld

1. The Problem

The reports of the transformation processes of selected research institutes in former socialist countries draw a picture of dramatic institutional change. Budget cuts, sometimes by half, staff reductions of similar magnitude, drastic shifts in the orientation of research: all of these changes brought about by the political turmoil of 1989 and after far surpass any experience made by the scientific community in the West. The political changes in the different countries affect science in very similar ways not least because of the similarity in organizational structures which the systems had acquired under Soviet influence since the end of World War II. It is all the more surprising to see how very differently scientists and science administrators in these countries face up to the challenges of institutional re-building and re-orientation of research.

In the following short essay I want to focus on one aspect revealed by the different analyses: the obvious difference in moods in which the changes are interpreted and which evidently determine the ways in which they are being dealt with. It is likely that these differences also indicate the chances of ultimate success. In order to explain these differences the attempt will be made to compare the most obvious and extreme cases, and to look for crucial variables. The possible precision of this exercise should not be overestimated, however. The reports are all second-hand, filtered through the interpretation of their authors. Also, the language is translated into English, often with considerable loss of representational fidelity, as is apparent in the mistakes. Finally, the choice of informants and institutions from which they come is completely outside our control, and their number is certainly too small to assure representativeness in any way. It should be clear, therefore, that most likely the reactions of the experts and researchers polled mirror more the *mood* and *outlook* of those interviewed rather than necessarily the actual situation. That is precisely what this essay is

R. Mayntz et al. (eds.), East European Academies in Transition, 141–149.

supposed to highlight. Thus, what follows is on the level of an impressionistic secondary analysis of the transformation studies.

I will deal with just three of these studies: those from Russia (Mindeli et al.), Belarus (Nesvetailov, this volume), and the Czech Republic (Provaznik et al., this volume). In some instances the material on international cooperation (Mirskaya, this volume) will also be used to complement the reports on the Russian institutes.

To grasp the differences in 'mood' and ways of coping with the institutional transformations I will distill from the material opinions and descriptions on several items:

1) Self-conceptions before and after 1989: how do scientists see their own role or that of their institution in the international scientific community? How do they judge their own quality?
2) What is their reaction to the organizational and intellectual changes? Do they see them as opportunities or as threats? Are they trying to defend the status quo or are they receptive, perhaps even enthusiastic, about changes?
3) How do their moods and attitudes relate to strategies of coping with the changes?

2. Self-Conceptions, Self-Deceptions

The survey of the Russian research institutes included one institute of economics, one in theoretical physics, one in radio-engineering and electronics, and one in biology (cf. Mindeli et al.) There are marked differences between these institutes which, apart from being affected differently by budget cuts and loss of personnel, have to do with the different scientific cultures to which they belong.

For example the physics institute gained a few researchers while the other three lost researchers: the economics institute lost ca. 40%. Another tendency, which is expressed differently across the institutes but common to all of them, is the aging of their scientific staff, in most cases going hand in hand with a 'brain drain', both internal and external. Opportunities to take up positions outside the research systems also differ for, say, economists and theoretical physicists. A further characteristic of the situation of scientists in Russia is that they often have to take up secondary jobs in order to get by economically. Clearly, those with competencies which are in demand outside academia have an advantage and are more flexible. But the downside effect is that their institutes are likely to suffer more.

The first indicator of the scientists' self-conception is their view of the quality of research at their institutes. The Russian researchers in the respective institutes were asked how they evaluated the level of research "in the past", i.e., 1990, and "now" (1994). The changes as reflected in the responses to these questions are dramatic indeed. Among the economists the percentage judging the level to be 'high' dropped from 67% to 5%, likewise the physicists from 100% (!) to 48%, the engineers from 50% to 5%, and the biologists from 42% to 8%.(Mindeli et al., supplement I, Fig. 3)

It is remarkable that the frame of reference of this self-assessment is a seemingly unrealistic image of Russian science before the collapse of the Soviet Union. Although the institutes are anonymous and there is no way of checking their claims to fame some doubts may be raised. Thus, the report states for the physics institute that it "keeps its leading position in basic research both in Russia and in the world", something hard to believe given the complaints of researchers about shortage of information, poor means of communication and the low technical state of instruments available (Mindeli et al., 39, 46). The radio engineering and electronics institute is described by two experts as "a leading centre in Russia and abroad" (ibid., 64), and similarly the biology institute qualifies, at least in the eyes of the experts queried, as "one of the leading centres in Russia and is very widely known abroad" (ibid., 90). Its present performance is qualified as "still keeping the leading position", an assessment that does not coincide well with the researchers' own evaluation of their level of work and the general personnel situation (ibid., 98 and supplement).

The description of the institutes' status in Belarus is conspicuously more modest. Again, one is a physics institute, the second is in engineering, and the third in biology. The situation is very similar to that described for Russia: budget cuts have caused reduction of personnel, the mean age of those who remain rises. The assessment of the future is pessimistic: while the quality has not fallen, in the eyes of one expert, progress in international science is so fast that chances to keep up are diminishing rapidly (Nesvetailov, this volume).

The engineering institute lost 60% of its production workers, and with a time lag an unspecified number of researchers has followed. Yet, the report notes an 'improvement of the socio-psychological climate in the institute (as) a major change that occurred in recent years' and characterizes the transformation period as one of 'active organizational innovations' (Nesvetailov, this volume). This is perhaps exemplified by a very conspicuous orientation to the international scene. Publications in Western countries increased from 8% in 1989 to 46% in 1994. At the same time the percentage of publications in countries of the former USSR dropped from 40% in 1990 to just 13% in 1994. Apparently this turn westward has also brought cooperative projects to the institute with exchanges of personnel, participation in international conferences, as well as infrastructural aid (Nesvetailov, this volume).

Finally, the biological institute is described as one of the 'major research institutions of biological sciences' which, in terms of international comparison, "could be estimated as average" (Nesvetailov, this volume). This institute, like all others, suffered heavy financial cutbacks and a decrease of personnel with the result that the quality level of research "appeared slightly lowered" (Nesvetailov, this volume). 78% of those interviewed consider the "intellectual core (of the institute) preserved so far", and believe that no "dramatic deterioration has occurred as yet" (Nesvetailov, this volume). More troubling to them seems the lack of funds and material equipment.

An important factor in the course of transformation seems to be that the previously existing relations to 'institutes of applied research and design offices' outside Belarus have almost all been severed since the breakdown of the USSR. At the same time "international cooperation has become the most dynamic process in the transformation", so that "abrupt expansion of scientific cooperation with Western research centres is accompanied by breaking contacts with the CIS countries and the former COMECON members" (Nesvetailov, this volume). The Belarus scientists seem to be fairly successful in attracting international funds, and, most important, do not observe a serious 'brain drain' of their young researchers (Nesvetailov, this volume).

Finally, I turn to the Czech institutes: one in physics, the second in chemistry, the third in philosophy. Although all three of them operate in the realm of basic research they represent quite different cases both in terms of their self conceptions and their initial (pre-transformation) status. It is striking how different the picture drawn is from that of the two preceding countries.

The institute of physics was the largest institute in the Academy framework before 1990. Its transformation began in 1990 in synchrony with that of the academy as a whole. The report gives little information on pre-1989 self-conceptions of the physicists except to say that the institute "was highly prestigious abroad even before the transformation" which could be interpreted as a rather abrupt break with the past (Provaznik et al., this volume). It is indicative of the mood that the 30% reduction in personnel is translated into an opportunity to get rid of inefficient researchers. That is possible for two reasons: the cuts were implemented on the basis of evaluations of individuals' research performance, and the 'efficiency' of research did, indeed, increase, at least as measured in terms of citations received by articles originating in the institute. While one has to be careful not to overinterpret citation data, especially since it is not known if (and unlikely that) they have been cleaned of self-citations, the average number of ca. 6 + citations per paper is noteworthy. More important as an indicator of the positive and reform minded mood is the readiness to implement evaluation procedures and to even accept evaluation by an internationally constituted committee. Apart from the instrumental effect of determining strengths and weaknesses of research, such a procedure undoubtedly signals both the selfassurance to face up to international critique and the integration into that community. This impression is supported by the institute's apparent success in securing international cooperative ties in several of its research fields (Provaznik et al., this volume).

The chemistry institute, originally in chemical engineering, then called the Institute of Theoretical Foundations of Chemical Technology (ICPF), was established in 1964. Under the impact of government autarchy policies it was gradually pushed into applied research and product development. In retrospect this led to the institute's losing its "credit as a renowned research centre" and by 1989 it is said to have merely survived. Thus, the transformation of the Academy were welcome for the institute as well, both in terms of change was personnel and a reorientation towards basic research in the framework of the new paradigm of chemical engineering (Provaznik et al., this volume). The goal which the institute set itself was to "Regain the international renown", a goal to be achieved by following the trends in the international community.

This rather self-critical assessment of the institute's pre-1989 performance has to be seen in conjunction with a dramatic turnover of personnel in 1990. Almost 50% of the entire staff had to leave, and, as the report states, "they were people of all categories of efficiency" (Provaznik et al., this volume). The age structure of the institute is also considered unfavorable. In other words, the transformation process as such is not, in retrospect, considered a complete success in terms of the plan but is in part conditioned by the political and economic circumstances. Yet, the institute is seen to reintegrate itself into the international scientific community and to thereby build its prestige abroad. Again, this self-perception is substantiated by a 70% share of its papers appearing in international journals, the winning of grants, and the participation in international co-operative projects. Like the physics institute, the ICPF has been evaluated by an international evaluation committee which ruled the transformation a success. Indeed, in spite of the staff reductions, the research output has increased, and the committee has noted that its standard meets those of the international community (Provaznik et al., this volume).

The third institute, that of philosophy, is a new foundation taking over from a predecessor in the Academy, the Institute of Philosophy and Sociology. Even though discontinuity is stressed, the organizational identity is apparently great enough to warrant comparisons between pre- and post-1989. In this case again, however, the stress on programmatic discontinuity and the complete reorientation (although now largely by virtue of individual choice) clearly indicates the ideological break with the past, which in philosophy and sociology is particularly evident.

It is not surprising, therefore, that the turnover of personnel between 1990 and 1993 is particularly high at 75%, and much higher than the actual net loss of ca. 25% (Provaznik et al., this volume). However, this institute, too, has undergone an evaluation which supports the self-conception that it has managed to attract high-quality staff. Those researchers working in the institute now were polled to judge the quality of those who had left in the period 1990-1993. Their self-image is that 70% of those who have left were of lower productivity, while 30% were considered to have been of standard quality. At the time of the poll the 18 researchers judged themselves as being one third top quality, and two third standard with no-one in the 'less productive' category. In their judgement of the factors which have contributed to this self-confident assessment the researchers mention "the elimination of ideological pressures and the possibility of conducting free research according to one's own choice" and the establishing of international contacts (Provaznik et al., this volume). It is not at all clear that the philosophers who were subject to ideological pressures more so than any other academic field have most decidedly moved towards the traditional pattern of individualistic research. Their self-assessment is corroborated by their reintegration into the international community of scholars, which they proudly note.

3. Comparative Observations on Success and Failure of Transformation

The pictures drawn by the scientists in the three countries about their relative success and failure in coping with the transformation of their institutes reveal marked differences. Russia and the Czech Republic are the two extremes of the cases which have been analyzed. Although the material does not allow a systematic comparison, there is enough evidence in both case studies to permit some conclusions, albeit with less methodological rigor. Also, the interview data from the individual institutes are supplemented by a (non-representative) survey of 300 scientists from 12 Academy institutes in Moscow carried out in 1994, which gauged both the prevailing mood and assessment of present conditions and future opportunities among Russian scientists, as well as the scope of international cooperation (Mirskaya, this volume).

It is very revealing to read the conclusions of the Czech report and Mirskaya's survey side by side in order to make sense of the very different approaches of the Russian and Czech scientists in dealing with the challenges posed by the transformation. The Czech report in an attempt to compare its own results singles out what the authors consider the important factors in accounting for successful transformation processes. As far as the three institutes which were under investigation are concerned they share certain strategies in dealing with the problems posed by the exigencies of transformation.

First, the significant reduction of personnel (50% for the Czech Academy as a whole) was carried out selectively on the basis of productivity and the promise of research fields.

Second, staff reductions were implemented on the basis of what the report terms an "objective evaluation of individual members". This process, operating with quantifiable criteria evidently came as a shock to researchers, but since then has been established as a recurrent procedure on both the individual and the institutional level (Provaznik et al., this volume).

Third, the system of funding shifted in emphasis from institutional funding to a competitive grant system which even applies to the larger share of researchers' salaries. The report makes clear that even though it is unknown whether this mechanism would guarantee a better quality of research in the long run "in the specific Czech conditions it was, and is, a decisive means of radical and quick changes" (Provaznik et al., this volume).

Fourth, there has been a determined and largely successful effort on the part of the institutes and the individual researchers to link up with the European and world communities in their respective fields. Their experience has been that "international co-operation is more productive when its organization is brought directly into the research-units themselves, and when the relevant international agreements ... are concluded at the level of the individual research teams" (Provaznik et al., this volume). This can be read as a reaction to the politically determined formalistic arrangements that characterized East-West science relations before 1989 and which were more often than not a form of political window dressing.

Fifth, the report lists "space for flexible and democratic search of optimal alternatives" (Provaznik et al., this volume). By this the authors circumscribe that the factors mentioned so far have been accepted by the community of researchers as a new value system, which allows each individual a higher degree of self-determination and an influence on matters of his/her concern, even though the price is more competition and insecurity. It does not matter if this reflects an uncritical enthusiasm with which the new 'Western' democratic and individualistic values are embraced. This is precisely the reflection of a mood which seems to prevail among Czech scientists and probably in wider intellectual circles in Czech society. This mood crystallizes into a sixth and final factor.

Sixth, and in my reading of the different analyses the crucial factor, is the attitude towards the future. The authors state that those institutes "which quickly eliminated old criteria of undemanding evaluation, anticipated future professional and social demands and, in a spirit of critical self-reflection actively initiated radical changes ... succeeded in their transformation incomparably better" than those institutes which followed a more cautious strategy (Provaznik et al., this volume). Those that tried to preserve the status quo and did not take into account the changes of a new market and new social demands for research, nor the international conditions to which they were now exposed, in fact failed to succeed in their efforts to preserve their research potential (Provaznik et al., this volume).

It is ironic that this diagnosis describes the Russian situation almost down to the last detail. There the dominant reaction to the necessary transformation appears to have been exactly the kind of defensive strategy which in the Czech case has proven to be unsuccessful. The attempts to defend the status quo were coupled with a submission to the changes occurring outside without any formulation of more precise goals.

How can it be explained that the reactions among scientists in two countries to very similar challenges are so different? Of course, there is probably no simple answer, especially since the cases presented undoubtedly hide variations within each of them. Nonetheless, I want to advance a thesis which can be supported by Mirskaya's survey data, among other things. My view is this: the prevailing mood among the Russian scientists is one of pessimism coupled with unrealistic self-assessment, which makes them unable to react more creatively to the threats from outside; the Czech scientists, by contrast, demonstrate an optimistic mood, but are much more self-critical and thus open to critical assessment from their peers.

Mirskaya found in her survey that 56% of her respondents thought that Russian science would be further destroyed, and only 28% answered on a more optimistic note that the crisis would be overcome. She states that the pessimistic mood is even growing, while the share of the more optimistic scholars is not diminishing (Mirskaya, this volume). Even more revealing are her data when she differentiates her sample into five groups on the basis of international reputation and contacts. (A = international reputation, publications in international journals, co-operation with colleagues in the West; all the way down to E = no international contacts, sometimes as a matter of principle, Mirskaya, this volumes). It turns out that group A, the elite, has a critical and largely realistic self-assessment but also the highest degree of job satisfaction. Thus, the scientists are most vulnerable to a destruction of their occupational infrastructure. Group B scientists, who are highly qualified even though not internationally renowned and who have the highest degree of outside jobs, are also the most pessimistic as to the prospects of a continuation of their work and its quality. Group D and E scientists, however, are characterized by a combination of unrealistic assessment of their institutes' and their own quality of work, which is in stark contrast to their actual performance. At the same time they propagate (most pronounced in E) a sense of catastrophism which seems to legitimate their own inactivity and defensiveness (Mirskaya, this volume).

Of course, Mirskaya's data, disregarding for a moment their methodological shortcomings, only mirror a stratification which can be found in similar fashion in every science system. By itself it would not explain the different reactions among Czech and Russian scientists. The additional factor is probably the relationship between the two countries prior to 1989 and their respective roles in the international scientific community. Russia, or rather the Soviet Union played the role of a world power and gave its weaker scientists the feeling of being part of that power even when this was not supported by their own achievements. For them the sudden loss of that position is very difficult to absorb. The temptation to hold on to the past and to hope for its recurrence is greater than for the Czech scientists. They probably have more to gain than to lose as far as their international stature in science is concerned. Thus, they embrace the 'new' values much more rapidly, perhaps even uncritically, as is demonstrated by their readiness to implement evaluation schemes and to open up to international competition and co-operation.

This interpretation of the general political transformation being translated down to the level of the psychology of the research communities in the former Soviet Union and its satellite countries is not counterintuitive and, thus, not very surprising. However, it leads to a seemingly paradoxical conclusion. If my interpretation is correct, the scientific community in Russia, and to an almost equal extent those in the other former member states of the Union, although emanating from a world power status, need more help and help more rapidly than those in the Eastern European states precisely because of the psychological disarray analyzed above. As Mirskaya correctly states at the end of her survey, however, such help must assume the form of mutual exchange and international co-operation among equals rather than just aid to the needy (Mirskaya, this volume). The dilemma is that such an integration into the international scientific community presupposes the kind of attitudes which the reports and surveys have shown to be lacking.

Reference

Mindeli, Levan, et al. (1995). *Transformation of academic science in the transition period: Case study results.* Moscow: Russian Academy of Sciences, Centre for Science Research and Statistics of the Ministry of Science and Technological Policy of the Russian Federation.

APPENDIX

R. Mayntz et al. (eds.), East European Academies in Transition, 151–155.

Part I: Belarus – Latvia

Countries	Research Institutions	Year of Founding	Personnel	Field of Research	Special Features of Transition
Belarus	institute of physics	1960's	1990: 320 1994: 322	basic research	–
	institute of engineering	–	1990: 1.434 1994: 584	applied research	–
	institute of biology	mid-1960's	1990: 334 1994: 248	basic research	–
Bulgaria	institute of nuclear research	1955	1990: 428 1994: 373	basic research	–
	institute of computer- and communication systems	1994	1994: 160	applied research	Created as a fusion of three former research institutes in the field of cybernetics, robotics, and informatics.
	institute of molecular biology	1960	1990: 158 1994: 143	basic & applied research	Brain drain of approx. 50 % of the research staff.
Czech Republic	institute of physics	1954	1990: 849 1994: 584	basic research	evaluation
	institute of chemistry	1964	1990: ~340 1995: 180	basic research	evaluation
	institute of philosophy	1990	1990: 120 1995: 86		Re-creation of the Philosophy Institute which was founded in the 60's and abolished by government in the 70's.

(continued next page)

(Appendix Part I continued)

Countries	Research Institutions	Year of Founding	Personnel	Field of Research	Special Features of Transition
Estonia	institute of chemical physics	1980	1990: 178 1994: 146	basic research	–
	institute of cybernetics	1960	1990: 188 1994: 114	basic & applied research	–
	institute of chemistry	1947	1990: 291 1994: 143	basic & applied research	Changes in the organizational structure
Hungary	institute of solid state research	1981	1990: 244 1994: 115	basic research	–
	institute of chemistry	1954	1990: 528 1994: 442	applied research	Created as a fusion of three former research institutes in the field of cybernetics, robotics, and informatics.
	institute of isotopes	1959	1990: – 1994: –	basic research	Radical change from applied to basic research
Latvia	institute of physics	1954	1990: 849 1994: 584	basic research	Specified research fields have been singled out of the institute and became institutes by their own.
	institute of polymer mechanics	1963	1990: 252 1995: 150	applied research	–
	institute of organic synthesis	1957	1990: 579 1995: 390	basic & applied research	–

Part II: Lithuania – Ukraine

Countries	Research Institutions	Year of Founding	Personnel	Field of Research	Special Features of Transition
Lithuania	institute of semi-conductor physics	1967	1990: 609 1994: 313	basic research	–
	institute of biochemistry	1967	1990: 293 1994: 206	applied research	–
	institute of Lithuanian language	1941	1990: 85 1994: 104		Re-creation as a "state institute" in 1992.
Poland	institute of physics	–	1991: 342 1994: 284	basic research	–
	institute of biology	1957	1991: 200 1994: 204	basic research	–
	institute of social sciences	1991	1991: 39 1994: 24		Founded after the liquidation of the former (marxist) institute of social sciences.
Romania	institute of physical chemistry	1949	1990: 227 1994: 291	basic research	–
	institute of theoretical informatics	1984	1989: 59 1995: 58	basic research	–
	institute of biochemistry	1950	1990: 53 1995: 51	basic research	This institute was liquidated in 1973 and re-created in 1990.

(continued next page)

(Appendix Part II continued)

Countries	Research Institutions	Year of Founding	Personnel	Field of Research	Special Features of Transition
Russia	institute of economics	1960's	–	basic research	–
	institute of theoretical physics	–	–	basic research, "Center of Superiority"	–
	institute of radio engineering and electronics	1950's	–	applied research	–
	institute of biology	–	–	basic research	–
Slovak Republic	institute of physics	1955	1990: 112 1994: 101	basic research	–
	institute of experimental pharmacology	1969	1990: 127 1994: ~60	applied research	–
	institute of ecobiology	1990	1990: 82 1994: 36	basic research	This institute is a re-creation of the "Institute of Experimental Biology and Ecology" which was founded in 1973.
Ukraine	institute of chemistry	1980's	1990: 568 1994: 492	basic research	–
	institute of physics	1945	1989: 1.004 1995: 883	basic research	–
	institute of informatics	1957	1990: 53 1995: 51	applied research	The institute is currently changing from applied to basic research.

SUBJECT INDEX

Academic
 Appreciation Committee 127
 global culture 51
Academy
 autonomy 125
 central bodies 128
 functions 131
 general statutes 128
 institutes
  defence industry 56
 leadership viii, 5, 52, 58
 legitimising function 133
 main funding source 129
 presidium 107
 process function 132
 protective function 133
 science autonomy viii, 86
 strategic functions 131
 symbolic function 134
 West European model 126
Academy of Sciences
 Belarussian 18, 35
  salaries in institutes 102
 Bulgarian 74, 86, 88
 Czech 8, 111
 East German viii, 7, 9, 67
 Estonian 20, 131
 Hungarian 9, 10, 49, 85, 111, 116, 135
 Latvian 116, 134
 Polish 41, 84, 111
 Romanian 18, 36, 38, 74, 115, 127, 135
 Royal Academy of Sciences (Great Britain) 109
 Russian 1, 8, 25, 37, 46, 107, 109
 Slovak 17, 36, 88
 Ukrainian 20, 41
 USSR 108
Academy-institute relations 125, 126, 129, 131, 134, 137–139
Academy-university relations 79, 80
Analytical Centre for Science and Industry Policy 96, 105
Autonomy vii, viii, xi, xvi, 3, 6, 7, 11, 24, 27, 50, 51, 83, 91, 110, 125, 126, 128, 129, 131, 132, 136–138
 of science vii, 6, 51
Baltic states 1, 41, 118, 126
Biologists 115, 116, 143
Brain-drain 10, 29, 71, 98, 122, 142, 144
Branch institutes viii, xii, 6, 52, 55, 68
CEE – Central and Eastern Europe vii, 108, 109, 111–114, 118, 119, 122, 123
 countries vii, 108, 110, 111, 113, 118, 119, 122, 123
researchers 114
Citation data 144
Co-operation
 international, preferred forms of 117
 scientific 108
COCOM – Coordinating Committee for Multilateral Export Controls 56
 embargo 56
Collaboration
 international types of 113
 intra-European 112
Collective research 122, 123
COMECON – Council for Mutual Economic Aid (1949-1991) 19, 43, 57, 107, 108, 110, 111, 144
Command economy 43
Consulting xi, 54
COPERNICUS – Community of Pan-European Research Network of International Centers and Universities in Sciences 116, 133
Coping xiv, xvi, 8, 9, 40, 45, 63, 67, 69–74, 76, 93, 146
 moods in ~141
 ways of ~ 142
 with scarcity 33
Council of research institutes 126
Countries investigated
 Austria 112
 Baltic Republics 6, 7, 11
 Belarus ix, 15, 18, 19, 35, 41, 51, 54–58, 70, 79, 86, 87, 89, 93, 97, 99, 101, 102, 112, 115, 127, 129, 130, 132, 137, 142–144, 152
 Belgium 112
 Bulgaria ix, 23, 36, 55, 57, 58, 71, 72, 74, 84, 86, 88, 96, 111, 115, 116, 117, 125, 127–129, 132, 133, 135–138, 152
 Canada 112
 China 112

Czech Republic ix, xvi, 5, 7, 8, 10–12, 15, 18, 21, 34, 35, 37, 38, 39, 41, 47, 53, 56, 69, 70–72, 76, 77, 84, 85, 88, 90, 91, 96, 98, 101, 104, 105, 111, 112, 114, 116–118, 127, 130, 132, 135, 142, 144, 146–148, 152
East Germany 3, 6, 7, 69
Estonia ix, 7, 15, 36, 84, 88, 89, 92, 126, 131, 137, 153
France 112, 115
Germany ix, 1, 3, 6, 7, 69, 112
Great Britain 112
Hungary ix, 7, 8, 10–12, 15, 18, 21–23, 34, 37, 44, 47, 49, 52, 82, 84, 85, 87–90, 104, 111, 112, 116, 118, 126–130, 132, 135, 137, 153
Italy 112
Japan 110, 112
Latvia ix, 15, 18, 19, 21, 23, 24, 83, 88–90, 96, 100, 126, 130, 131, 134, 152, 153
Lithuania ix, 18, 71, 72, 76, 83, 85, 88, 89, 96, 100, 101, 115–117, 126, 128, 129, 131, 154
Poland ix, 7, 9, 15, 16, 18, 21–23, 34, 35, 37, 38, 41, 44, 47, 49, 55, 57, 69, 73, 75, 76, 83–85, 88, 89, 96, 97, 100, 102–105, 108, 111–116, 118, 122, 126–129, 132, 135, 137, 154
Republic of South Africa 112
Romania ix, 6, 18, 21, 23, 36, 38, 51, 52, 55, 84, 86, 88, 98, 100–102, 106, 111, 112, 115, 116, 118, 127, 129, 132, 135, 137, 138, 154
Russia ix, xiv, xvi, 1, 2, 6, 8, 12, 15, 19, 25, 26, 36–38, 41, 42, 46, 51, 52, 54–56, 58, 59, 68, 70–73, 75, 76, 79, 85, 87, 89, 92, 93, 96, 98–102, 104–107, 109, 111, 112, 116–119, 122, 123, 129, 135, 138, 142, 143, 146–148, 155
Slovak Republic 7, 10–12, 15, 17, 21–23, 36, 47, 53, 84, 86, 88, 89, 98, 100, 101, 104, 111–114, 116, 117, 127, 130, 132, 135, 155
Soviet Russia 1
Soviet Union 1, 7, 15, 18, 42, 83, 107, 108, 111, 114, 116–118, 134, 143, 148
Spain 112
Sweden 112
Switzerland 112
Ukraine ix, 15, 16, 20, 35, 36, 41, 42, 51, 54–56, 60, 73, 74, 79, 85, 87, 89, 91, 93, 97, 100, 104, 128, 129, 133, 138, 154, 155
USA 108, 112, 116
EC – European Community 19, 58, 116
programmes 117
Economic crisis vii, 8, 15, 18, 45, 53, 55, 63, 91, 98
Economists 142
Engineers 100, 143
EUREKA – Europen Research Coordination Agency 116
European Community – EC 116
Evaluation 6, 8–10, 17–19, 21, 28, 37, 42, 44, 45, 47, 51, 65, 68, 69, 76, 78, 118, 127, 135, 137, 143–148, 152
Extra-scientific
usefulness 71, 72, 75–77
users viii, xv, 65, 67
Failure of organization 64
Finances vii, ix, xi, xiv, 21, 23, 42, 45, 52, 71, 86, 105
Financial
aspects xiv
means 42, 65, 66
reductions xiv, 66
resources vii, 4, 27, 41, 42, 65, 67, 68, 70–72, 76, 77, 84, 91, 130
Financing
financial situation 59
government ~ 49
state ~ 59
Fiscal crisis 53
Funding 7, 10, 13–16, 18–21, 25, 27, 28, 41, 42, 45, 46, 52, 53, 58, 59, 65, 68, 70, 71, 76, 80, 81, 84, 86, 88–91, 102, 105, 110, 111, 115, 117, 127, 133, 146
grant ~ 88
institutional ~ 20, 53, 65, 70, 76, 146
project ~ 42, 65, 71
Funds
institutional 25, 58, 66, 69, 70
project 66, 69, 70, 74
G. Soros Foundation 114, 117
GDP – Gross Domestic Product 15, 41, 43
Government actors 52, 72

Grant(s) viii, xi–xiii, 4–6, 9, 10, 19–27, 29, 30, 35, 42, 52, 68, 84, 88, 91, 96, 102, 109, 113, 114, 117–120, 123, 130, 132, 134, 145, 146
  funding 88
Humanities x, 37, 64, 75–78, 107, 109, 110, 134
ICSOPRU (Subdivision of UNESCO) – International Comparative Study on the Organization, Productivity and Effectivity of Research Teams and their Institutions 94, 95
Industry vii, viii, x, 6, 7, 15, 16, 19, 20, 35, 42, 45, 50, 53, 55–58, 72, 74
Infrastructural 22, 23, 27, 71, 129, 138, 143
Infrastructure xi, xiv, 2, 8, 10, 23, 25, 55, 65–69, 80, 89, 92, 130, 148
Institute
  activity spectrum x
  autonomy 128, 138
  financial resources vii, 4, 27, 41, 42, 65, 67, 68, 70–72, 76, 84, 91, 130
  financial situation 59
  infrastructure xi, xiv, 2, 8, 10, 23, 25, 55, 65–69, 80, 89, 92, 130, 148
  level of performance 65
  mode of financing 51
  relations with
    Academies 125, 126, 129, 131, 134, 137–139
    institutes of applied Research xii
    political actors xii
    the universities xii
    users of research results xiii
Institutes
  Belarussian
    institute of biology 58, 70, 137
      applied research 152
      institutional funds 58
    institute of engineering 19, 51, 99
    institute of physics 18, 54, 55, 58, 97, 101, 115, 130, 152
  Bulgarian
    institute of computer and communication systems 55, 58, 74
    institute of molecular biology 71, 86, 96, 152
    institute of nuclear research 57, 72, 88, 111
  Czech
    institute of chemistry 96, 105, 118
    institute of natural sciences 37
    institute of philosophy 37, 70–72, 76, 77, 152
    institute of physics 53, 98, 101, 104, 116, 152
    institute of social sciences and humanities 37
  Estonian
    institute of chemical physics 102, 153
    institute of chemistry 18, 56, 116, 137, 153
    institute of cybernetics 20, 153
  Hungarian
    institute of chemistry 97, 115, 116, 118, 153
    institute of isotopes 102, 103, 153
    institute of solid state research 93, 96, 153
  Latvian
    institute of organic synthesis 70, 83, 100, 102, 153
    institute of physics 70, 86, 153
    institute of polymer mechanics 68, 75, 100, 153
    Latvian language institute 37
  Lithuanian
    institute of biochemistry 100, 101, 116, 154
    institute of Lithuanian language 18, 71, 72, 76, 154
    institute of semi-conductor physics 96
  Polish
    institute of biology 35, 38, 57, 69, 73, 96, 103, 105, 114, 116, 154
    institute of physics 55, 69, 75, 96, 97, 100, 102, 104, 115, 118, 154
    institute of social sciences 69, 73, 76, 137, 154
  Romanian
    institute of biochemistry 36, 38, 101, 102, 138, 154
    institute of physical chemistry 36, 51, 55, 98, 118, 154
    institute of theoretical informatics 36, 52, 101, 154
  Russian
    institute of biology 25, 26, 54, 70, 72, 73, 99, 101, 104, 155

institute of economics 46, 73, 76, 93, 99, 101, 155
institute of radio engineering and electronics 25, 26, 58, 59, 70, 98, 101, 102, 155
institute of theoretical physics 25, 38, 52, 59, 68, 71, 75, 155
Slovak
institute of ecobiology 69, 88, 155
institute of experimental pharmacology 17, 100, 104, 155
institute of pharmacology 69
institute of physics 53
institute of physics 69, 98, 101, 155
Ukrainian
institute of chemistry 17, 20, 35, 54, 97, 155
institute of informatics 20, 36, 54, 55, 60, 74, 100, 155
institute of physics 55, 93, 97, 128, 155
Institutional
changes 4, 7
financing xi, 23, 24, 67, 71
funding 19, 20, 53, 65, 70, 76, 146
funds 25, 58, 66, 69, 70
International
co-operation, preferred forms of ~ 117
collaboration, types of ~ 113
competition 77, 148
contacts 38, 65, 108–110, 112, 114, 116, 119, 139, 146, 148
networks xvi, 68
relations xvi, 107, 109–114
science x, 82, 109, 117, 122, 143
science relations 109
scientific co-operation 102
scientific community 45, 53, 108, 109, 113, 118, 142, 145, 148, 149
Science Foundation 102, 105, 110, 123
Joint
project 23, 26, 81, 82, 85, 88–90, 116, 117, 119, 122, 128
research xii, 55, 81, 89, 90, 92, 103, 113, 116
Learned society(ies) 1, 2, 5, 6, 7, 10, 11, 83, 126
Mobility 35, 42, 81, 84, 91, 100, 105
Mood(s) xvi, 141, 142, 144, 146–148
differences in 142
moods in coping 141
National academies of sciences viii, ix, xiv, 1, 63, 80 (see also *Academy of sciences'*
National Science Foundation (NSF) 12, 117
Natural sciences xv, 1, 37, 38, 64, 72, 75–77, 94, 134
OECD –Organization for Economic Cooperation and Development 2, 10, 12, 43, 45, 47, 81
Organizational structures xi
autonomy of researchers xi
decision-making procedures xi
Over-staffing 127
PECO –Pays d'Europe Centrale et Orientale 102, 116, 133
Personnel ix, xi, xii, xv, 2, 5, 8, 9, 19, 23, 25, 33, 43, 65–71, 83, 84, 86, 91, 93, 95– 100, 102–105, 117, 125, 135, 142–145, 152–155
age structure xi, 93, 95, 96, 100, 105, 145
aging ~ xv, 66, 68
reduction of xiv, 65, 96, 143, 146
turnover of 145
PHARE – Programme for Harmonization of the ATM (Air Traffic Management) Research in Eurocontrol 58, 116, 133
Philosophy 2, 37, 70–72, 75–77, 144, 145, 152
Physicists 111, 115–117, 142, 144
theoretical 142
Political actors xii, xiii, 11, 51, 53, 57, 58, 71
Postgraduate studies 81, 87
Programme for the development of fundamental studies 127
Publications
distribution by language 118
in foreign journals 118, 119
increasing number of 118
R&D – Research and Development 2, 4, 6, 10, 15–17, 19, 21, 34, 35, 37, 38, 40, 43–45, 49, 52, 54, 73, 79, 81–87, 90, 91, 100
Research
aging employees 65
aging personnel 66
aging staff 93

applied viii, xii, xv, 2, 6, 20, 21, 24, 40, 49–61, 64, 71, 73–75, 77, 144, 152–155
finalized viii
basic viii, x, xv, 1, 2, 6, 10, 11, 20, 23, 49, 50–61, 64, 71, 73–76, 78, 94, 97, 131–133, 138, 143–145, 152–155
basic-oriented xiii
commercialisation 136
contract ~ vii, xiii, 6, 7, 17
contracts xi, 19, 27, 54, 57, 58, 66
freedom of institutes 49
fundamental 49, 57, 80, 87, 131, 136, 139
funds 14, 25, 130
grants viii, xi, xii, xiii, 5, 6, 9, 10, 19–27, 29, 30, 35, 52, 68, 84, 96, 102, 109, 113, 114, 117–120, 123, 134, 145
high quality xv
infrastructure 8, 10, 65, 80, 89, 92
joint ~ xii, 55, 81, 89, 90, 92, 103, 113, 116
low quality xv
mission-oriented x
organization xiv, 1–3, 5–7, 10
performance 51, 67, 68, 144
plans 5
policy 24, 50, 51, 53, 57, 68, 72, 75, 77, 78, 101, 126
socialist viii
problem-oriented 1, 11
projects 19, 20, 42, 43, 68, 69, 80, 81, 90, 99
project funds 66, 69, 70, 71, 74
pure x
results x, xiii, xv, 2, 43, 53–57, 65, 71, 82, 93, 94, 136
external demand in industry 50
self-regulation 63
staff 8, 23, 34, 36–38, 40–44, 66, 85, 86, 95, 103, 152
topics x, xi, 4, 10, 54, 55, 70, 71, 76, 128, 136
types of xv, 1, 49, 52, 58, 59, 60, 64, 67, 71
Researchers
aging 68
income xi
young ~ xii, xv, 67, 68, 71, 73, 95–97, 99, 100–105, 144
drain of ~ abroad 97
Resource scarcity viii
Resources vii, xvi, 4–6, 8, 18, 24, 27–31, 41–44, 53, 63, 65–68, 70–72, 76, 77, 81, 84, 91, 127, 129, 130, 136, 138
S&T – Science and Technology 2, 92, 96
policies 92
Scarcity of financial resources vii
Science and engineering indicators 108, 123
Science Citation Index 108
Science policy 3, 5, 6, 22, 59, 101, 129, 133, 137
Scientific
aging of ~ staff 142
community xiii, 10, 35, 44, 82, 83, 87, 95, 118, 121–123, 128, 141, 148
international xiii
stratification of 121
quality 64, 66–71, 75, 77, 78
in the humanities 64
in the natural sciences 64
in the social sciences 64
school 68, 99
traditions 99
Social sciences 37, 64, 75–78, 134, 135, 137, 154
Socialist
ideology 75
research policy viii
Soviet type of scientific organisation 80
fundamental research 80
postgraduate studies 81
Spheres
business 60
political 60
Staff numbers 28, 35, 36, 44, 46, 84
Strategy(ies)
coping ~ 9
coping with financial crisis 31
leadership viii, 5, 10, 52, 58, 65
survival 34, 45, 53, 60, 125, 130, 136
Success of organization 64
Teaching viii, xi, xii, 9, 69, 73, 80–82, 84–86, 88, 90, 92
TEMPUS – Trans-European Mobility Programme for University Studies 58, 102, 116, 133
TESSA – Training And Education In Strategically Significant Areas 117

Universities viii, ix, xii, xv, 1, 5, 7, 11, 13, 40, 42, 47, 79–92, 96–100, 103, 105, 106, 125, 130, 132, 141
branch institutes viii, xii, 6, 52, 55
branch sectors ix, 52
UNESCO – United Nations Educational, Scientific and Cultural Organization 94, 106
USSR
Perestroika 109
Western countries 11, 37, 50, 52, 71, 73, 143
Western restrictions on transferring advanced technology to socialist countries 74
Western sources 53, 68, 69, 75
funding agencies 53, 68, 69, 75

AUTHOR INDEX

Andrews, Frank M. 106
Braun, Dietmar 12
Bush, Vannevar 3, 12
Couderc, Marie-Laure 6, 12
Dacin, Anca 6
Darvas, György ix, xv, 9, 49, 73
Debackere, K. 106
Filácek, Adolf xiv, 6, 17, 28, 33
Gaponenko, Nadezhda 81, 92
Jablecka, Julita ix, xiv, 4, 13, 36, 41, 42, 49, 66
Kocka, Jürgen 3, 12
Kuhn, Thomas 94, 105
Lehmann, N. C. 93, 94, 105
Levin Stephan 105
Levin, Paula, 105
Levin, Sharon 105
Loudin, Jiří 33
Machleidt, Petr ix, xiv, 6, 17, 28, 33
Mayntz, Renate ix, xiv, xvi, 1, 2, 7, 12, 63, 78
Merton, Robert K. 93, 106
Messeri, Peter 95, 106
Mikulinski, S. 99, 106
Mindeli, Levan ix, xiv, 142–149
Mirskaya, Elena ix, xvi, 37, 45, 66, 103, 107, 109, 123, 142, 146–149
Müller, Karel 10, 12
Nadiraschwili, Alexander ix, xiv
Nesvetailov, Gennady A. ix, xv, 10, 29, 38, 66, 91, 93, 97, 106, 142, 143
Pelz, Donald C. 106
Provaznik, Stanislav xiv, 2, 17, 28, 33, 66, 142, 144, 146
Radulescu, Dan 100, 103, 106
Rappa, M. 95, 106
Sandu, Steliana 6
Sarmir, Eduard ix, xv, 4, 42, 79, 96
Schimank, Uwe vii, ix, xv, xvi, 8, 12, 38, 63, 73, 78, 92
Schweitzer, G. E. 103, 106
Simeonova, Kostadinka ix, xvi, 4, 52, 83, 125
Simonton, D. K. 93, 106
Stankiewicz, R. 95, 106
Stewart, John A. 95, 106
Stucke, Andreas 8, 12
Weingart, Peter ix, xvi, 63, 78, 141
Wolf, Hans-Georg 9, 12, 67, 69, 76, 78
Zajac, Stefan xv, 42, 79, 96
Zaman, Gheorghe ix, 6
Zuckerman, H. 93, 106

LIST OF CONTRIBUTORS

Darvas, György, Dr., Institute for Research Organization of the Hungarian Academy of Sciences, P.O. Box 4, H-1361 Budapest, Hungary.

Filácek, Adolf, Dr., Centre for Science, Technology, and Society Studies at the Institute of Philosophy of the Academy of Sciences, Jilksá 1, 110 00 Prague 1, Czech Republic.

Jablecka, Julita, Dr., Centre for Science Policy and Higher Education, University of Warsaw, N. Swiat 69, PL-00-046 Warsaw, Poland.

Machleidt, Petr, Dr., Centre for Science, Technology, and Society Studies at the Institute of Philosophy of the Academy of Sciences, Jilksá 1, 110 00 Prague 1, Czech Republic.

Mayntz, Renate, Prof. Dr., Max Planck Institute for the Study of Societies, Lothringer Strasse 78, D-50677 Köln, Federal Republic of Germany.

Mirskaya, Elena, Prof. Dr., Sociology of Science Department, Institute of the History of Science and Technology, Russian Academy of Sciences, 1/5 Staropansky per., 103 012 Moscow, Russia.

Nesvetailov, Gennady, Prof. Dr., CIS, Academy of Sciences of Belarus, Surganova str. 1, bldg. 2, 220 600 Minsk, Belarus.

Provaznik, Stanislaw, Dr., Centre for Science, Technology, and Society Studies at the Institute of Philosophy of the Academy of Sciences, Jilksá 1, 110 00 Prague 1, Czech Republic.

Sarmir, Eduard, Dr., Slovenská Akadémia Vied Prognostický ústav, Sancová 56, 81 105 Bratislava, Slovakia.

Schimank, Uwe, Prof. Dr., Fachbereich ESGW, Soziologie, Fernuniversität Hagen, Feithstrasse 140, D-58084 Hagen, Federal Republic of Germany.

Simeonova, Kostadinka, Prof. Dr., Centre for Science of Science, Bulgarian Academy of Sciences, ul. Serdika 4, Sofa-1000, Bulgaria.

Weingart, Peter, Prof. Dr., Faculty of Sociology, University of Bielefeld, P.O. Box 10 01 31, D-33615 Bielefeld, Federal Republic of Germany.

Zajac, Stefan, Dr., Slovenská Akadémia Vied Prognostický ústav, Sancová 56, 81 105 Bratislava, Slovakia.

Batch number: 10823959

Printed by Printforce, the Netherlands